Pervasive Displays

Understanding the Future of Digital Signage

Synthesis Lectures on Mobile and Pervasive Computing

Editor
Mahadev Satyanarayanan, *Carnegie Mellon University*

Mobile computing and pervasive computing represent major evolutionary steps in distributed systems, a line of research and development that dates back to the mid-1970s. Although many basic principles of distributed system design continue to apply, four key constraints of mobility have forced the development of specialized techniques. These include: unpredictable variation in network quality, lowered trust and robustness of mobile elements, limitations on local resources imposed by weight and size constraints, and concern for battery power consumption. Beyond mobile computing lies pervasive (or ubiquitous) computing, whose essence is the creation of environments saturated with computing and communication, yet gracefully integrated with human users. A rich collection of topics lies at the intersections of mobile and pervasive computing with many other areas of computer science.

Location Systems: An Introduction to the Technology Behind Location Awareness
Anthony LaMarca and Eyal de Lara
2008

Replicated Data Management for Mobile Computing
Douglas B. Terry
2008

Application Design for Wearable Computing
Dan Siewiorek, Asim Smailagic, and Thad Starner
2008

Controlling Energy Demand in Mobile Computing Systems
Carla Schlatter Ellis
2007

RFID Explained: A Primer on Radio Frequency Identification Technologies
Roy Want
2006

Pervasive Displays: Understanding the Future of Digital Signage
Nigel Davies, Sarah Clinch, and Florian Alt

ISBN: 978-3-031-01356-0   paperback
ISBN: 978-3-031-02484-9   ebook

DOI 10.1007/978-3-031-02484-9

A Publication in the Springer series
SYNTHESIS LECTURES ON MOBILE AND PERVASIVE COMPUTING
Series ISSN: 1933-9011 print   1933-902X ebook

Lecture #11
Series Editor: Mahadev Satyanarayanan, *Carnegie Mellon University*

First Edition
10 9 8 7 6 5 4 3 2 1

Pervasive Displays

Understanding the Future of Digital Signage

Nigel Davies
Lancaster University, UK

Sarah Clinch
Lancaster University, UK

Florian Alt
University of Munich (LMU)

SYNTHESIS LECTURES ON MOBILE AND PERVASIVE COMPUTING #11

ABSTRACT

Fueled by falling display hardware costs and rising demand, digital signage and pervasive displays are becoming ever more ubiquitous. Such systems have traditionally been used for advertising and information dissemination, with digital signage commonplace in shopping malls, airports and public spaces. While advertising and broadcasting announcements remain important applications, developments in sensing and interaction technologies are enabling entirely new classes of display applications that tailor content to the situation and audience of the display. As a result, signage systems are beginning to transition from simple broadcast systems to rich platforms for communication and interaction.

In this lecture we provide an introduction to this emerging field for researchers and practitioners interested in creating state-of-the-art pervasive display systems. We begin by describing the history of pervasive display research, providing illustrations of key systems, from pioneering work on supporting collaboration to contemporary systems designed for personalized information delivery. We then consider what the near future might hold for display networks—describing a series of compelling applications that are being postulated for future display networks. Creating such systems raises a wide range of challenges and requires designers to make a series of important trade-offs. We dedicate four chapters to key aspects of pervasive display design: audience engagement, display interaction, system software and system evaluation. These chapters provide an overview of current thinking in each area. Finally, we present a series of case studies of display systems and our concluding remarks.

KEYWORDS

pervasive displays, digital signage, ubiquitous computing

Contents

List of Figures

Acknowledgments

The research leading to these results has been conducted as part of the e-Campus and PD-NET projects. The PD-NET project acknowledges the financial support of the Future and Emerging Technologies (FET) programme within the Seventh Framework Programme for Research of the European Commission, under FET-Open grant number: 244011. We would like to thank all the staff and students that have contributed to the e-Campus project at Lancaster University.

Nigel Davies, Sarah Clinch, and Florian Alt
December 2013

CHAPTER 1

Introduction

1.1   PERVASIVE DISPLAYS AND DIGITAL SIGNAGE

No one can have failed to notice the increasing proliferation of digital signage. As hardware costs have decreased, the prevalence of digital displays in public spaces has grown considerably; displays of varying sizes, shapes and forms are now commonly seen in everyday spaces. Train stations and airports often have digital arrival and departure boards (indeed, finding an old-style analog display is more of a challenge), while conference venues and hotel lobbies use digital displays to share information on current and forthcoming events. Displays are also becoming prevalent in public spaces such as city squares and the workplace—replacing traditional notice boards and signage.

As compared to many other communications media, digital signage has a number of interesting characteristics that makes it extremely popular, especially with advertisers:

Push-based distribution.   Viewers do not need to make an active decision to access content on a typical digital sign. Rather, the content that is shown is often content that the users themselves would not actively seek out but that others wish them to see—advertisements being of course the canonical example. However, this ability to push information to viewers has applications beyond advertisements—for example in areas such as information dissemination or emergency announcements.

FIGURE 1.1: Examples of public displays. (Courtesy of Nemanja Memarovic, USI).

Context-specific content. The fact that signs are physically embedded in the world means that the content they show can (and indeed should) be related to their physical context. This is in contrast to many other communications media in which information is often presented out of context—typically when the user is unable to actually act on the message. A notable exception to this is of course web-advertising, which is deliberately designed to be context-based and by providing clickable links allows users to act on messages immediately.

Multimedia content. Unlike their traditional paper counterparts, digital signs can display a wide range of media types. In particular, digital displays are able to show moving images that are particularly effective at attracting viewers' attention. This provides new opportunities for engaging content.

Easy to update and efficient use of physical space. Traditional signs require a worker to physically visit them to update their content. In contrast, digital signs can be updated many times a day, from thousands of miles away and for minimal cost. This enables digital sign screen real estate to be shared at a very fine temporal granularity, enabling more efficient use of the potentially very valuable physical space occupied by the display.

Digital displays offer the potential to enhance our public spaces. Digital signage improves over traditional notices by facilitating frequent, timely updates, increasing accuracy and enabling provision of highly dynamic information that would not otherwise have been made available. Such displays also offer new possibilities for improving the aesthetics of a space by allowing the presentation of digital artwork, video or other media. Furthermore, the introduction of personalized, interactive content on public displays has the potential to promote viewer engagement with the space and to encourage social interaction within the space.

Displays also have a rich research heritage. Early work such as [18, 85, 87] focused on using displays as tools for communication and collaboration—principally driven by interest in CSCW. More recently, researchers have begun to investigate a much wider range of applications for display networks (e.g., as a replacement for a classified notice board [2]).

In terms of research communities, the study of displays spans many disciplines. Technical work is often carried out by the graphics communities (especially research focused on creating video walls) and the ubiquitous computing community. Indeed, Weiser's original ubiquitous computing vision included large displays (yard-scale devices) as an integral part of the world he foresaw. In addition to this technical work, pervasive displays are also studied in the social sciences, and they attract significant interest from the commercial sector.

One consequence of many communities working on this topic is that a number of different terms have emerged for essentially the same concept. In this lecture we use, interchangeably, *digital signage*, *pervasive displays* and *public displays* to all mean the same thing, i.e., collections of digital displays deployed in public or semi-public spaces.

Of course, digital signs are not without their shortcomings. In particular, compared to web-based content it is hard to understand how many people actually see any given digital sign (and hence supporting business models based around the number of impressions or views of a piece of content is difficult). Furthermore, there is no signage equivalent of a user "clicking through" on a piece of content. This makes encouraging and, more specifically, tracking user actions extremely challenging in the signage world. We will return to these issues in later sections on audience engagement and interaction.

A further difficulty with digital signage is in actually encouraging viewers to see and interact with content. Indeed, the creation and distribution of engaging content is key to realizing the potential offered by the installation of digital displays in public spaces. Currently, despite their prevalence, public engagement with digital displays in their environment is typically very low—viewers have become skilled at ignoring them [138]. Research has shown that many passersby look at public displays for less than two seconds [75, 86]. Typical commercial display deployments are dominated by generalized advertising content that is broadcast regardless of the current user set and rarely tailored to individuals viewing the display. As a result, viewers tend to assume that the content being shown is not relevant to them and exhibit what is called "display blindness" [138]. To compensate, content creators often simply add more motion or color to their displays in order to try and attract the attention of viewers, but this "arms race" is clearly unproductive for both sides—viewers are most likely ignoring displays simply because they are uninterested in the content.

The key challenge for designers of future display systems is to create innovative new systems that deliver real value to passersby. As new display and sensing technologies become available to support this innovation, we believe that such innovation will occur and that digital signage will emerge as one of the core components of our future computing infrastructure.

1.2   DISPLAY HARDWARE AND CHARACTERISTICS

A wide range of display hardware is available (and emerging) for creating pervasive display networks. Early public signage was often based on split-flap displays in which characters are displayed using a series of flaps that are rotated mechanically to form the display (see Chapter 3). Such displays were common in airports and train stations for many years. However, it was the emergence of relatively large and cheap LCD displays that has really fueled the growth in digital signage. LCD displays are typically 40" or larger and can be installed in many locations, offering an easy route to high-density signage. It is worth noting at this point that LCD displays designed for use in digital signage and pervasive display applications are typically designed to have a much higher specification than

FIGURE 1.2: Independently addressable lighting elements being used as a display.

standard domestic displays. Improvements as compared to domestic displays often include support for extended periods of continuous operation (up to and including 24*7), longer screen life, higher resistance to screen burn-in, ability to be mounted in landscape or portrait orientation and remote control of the display via an RS232 or similar interface. Modern LCD signage products may also include additional features such as automatically entering power-saving modes when no viewers are present in front of the display. LCD signs work well in most indoor conditions, and special-purpose high-contrast environmentally sealed units are available for deployments outdoors.

One key problem with using traditional LCD technology is the inherent limitation on the physical size of the display. To overcome this there has been a great deal of work on creating "video walls" from multiple displays. In such systems displays are tiled to form large display surfaces. A video wall may be driven by a single computer with appropriate hardware support or multiple computers with software being used to coordinate the display of relevant content. An example of work in this area is the SAGE™: Scalable Adaptive Graphics Environment that provides an open-source platform for displaying rich information.

While standard LCD panels and associated video walls form the vast majority of displays in use today, there are interesting alternatives such as the use of projectors. This type of technology can be very effective at providing large-scale displays at relatively low cost. However, projectors typically do not work well in daylight, and the noise caused by the fans required to cool projectors means that they are not well suited to regular signage duties. Indeed, while they are cost effective at producing large displays, they also tend not to support long periods of continuous operation without regular maintenance.

Innovation in the area of display hardware continues at a rapid pace. For example, recent research such as [35] has shown how multiple, independently addressable lighting elements can be automatically composed to form 3D displays. Such technologies are particularly interesting when considering large displays such as building or exhibition lighting.

Of course not all displays have to be mounted in the environment—many people already carry displays with them in the form of a smartphone or tablet computer. However, such displays do not really fall within the scope of this lecture, as they are not traditionally considered to be public signage. Where these devices are interesting to us is as adjuncts to larger public displays. For example, personal mobile devices could be used to display privacy-sensitive information that supplements less sensitive information displayed publicly. The use of mobile devices in conjunction with large displays is a topic we return to in Chapter 5. Perhaps the most exciting developments in this area is the emergence of viable wearable devices such as Google Glass™.

Using wearable displays, the potential for augmenting the physical space with highly personalized content is transformed. We do not consider Google Glass in detail in this lecture, but many of the ideas of personalization and pervasive content that we discuss in the context of pervasive displays have clear parallels when considering wearable computers.

1.3   LECTURE OVERVIEW

In this lecture we provide an introduction to the field of pervasive displays for researchers and practitioners interested in creating state-of-the-art display systems. In Chapter 2 we present a chronological view of the history of pervasive display research, providing illustrations of key systems from pioneering work on supporting collaboration to contemporary systems designed for personalized information delivery. In Chapter 3 we consider what the near future might hold for display networks—describing a series of compelling applications that are being postulated for future display networks.

Creating pervasive display systems raises a wide range of challenges and requires designers to make a series of important trade-offs, and we dedicate four chapters to key aspects of pervasive display design: audience engagement, display interaction, system software and system evaluation. In Chapter 4 we describe current models that attempt to allow designers to understand how viewers and passersby might engage with displays. This work covers a wide range of display types and applications. Once a user has become engaged, most future systems will need to support interaction with the display. In Chapter 5 we describe the state-of-the-art in display interaction—focusing mainly on interaction techniques that involve a mobile device but also including a brief survey of other interaction techniques, including gesture-based displays. The software that underpins pervasive display networks is described in Chapter 6.

Evaluation of systems is always important to researchers, and in Chapter 7 we describe a series of evaluation techniques that can be used in the context of pervasive display research.

To demonstrate some of the practical implications of the tools and techniques we have described, we present in Chapter 8 two case studies: e-Campus and Digifieds. Finally, Chapter 9 presents our conclusions.

CHAPTER 2

The History of Pervasive Displays

2.1 OVERVIEW

Pervasive displays and digital signage have a rich research heritage dating back over 30 years. During the course of this lecture we will describe many systems when we discuss specific aspects of pervasive displays, such as interaction or systems software. However, if we were to only present systems in this fashion, we run the risk of not conveying an overall view of how the field has developed over the long term. As a result, in this chapter we focus on providing a high-level summary of pervasive display research in roughly chronological order, clustering systems according to the major research themes of the time, i.e.,:

1. 1980s–1990s: Displays as Media Links
2. Mid- to late-1990s: Ambient and Wearable Displays
3. Early 2000s: Supporting the Workplace
4. Early to mid-2000s: Promoting Social Interaction and Community
5. Late 2000s: Long-Lived Deployments

Of course not all research on a specific topic fits into the relevant time frame so, where appropriate, we also provide pointers to follow-up work. We hope that this chapter provides the reader with a simple way of gaining a broad appreciation of the research that as been conducted in the field of pervasive displays.

2.2 1980s–1990s: DISPLAYS AS MEDIA LINKS

The use of digital displays in public spaces first emerged as a significant topic of research in the 1980s. Early work took the form of "media links," using video and audio links to connect together physically separate spaces. For example, Kit Galloway and Sherrie Rabinowitz created the "Hole-In-Space" [69], a three-day art installation in November 1980. The installation featured two large back-projected displays (plus speakers and cameras) installed in sidewalk-facing windows of the Lincoln Center for the Performing Arts in New York City and "The Broadway" department store in Los Angeles. A satellite link between the two cities allowed the creation of virtual windows in which the video feed of New York was shown on the screen in L.A. and the video from L.A. in New York.

Later projects in this period experimented with longer-lived connections, using media links to facilitate interactions between workers in multi-site research institutions. The Xerox PARC Media Spaces [18, 70] connected researchers at sites in Palo Alto and Portland by providing steerable video and audio links in the "common area" of each site. The media links ran 24 hours a day, seven days a week for over two years, finishing only when the offices in Portland closed. While originally intended to support formal meetings, the majority of interactions over the links were chance encounters lasting for less than five minutes. A similar system at Bellcore Labs, the VideoWindow [63], connected researchers on two different floors of the building using large projected displays in common areas.

While the majority of the work took place in the 1980s and 1990s, a second wave of media links on large public displays was developed during the early 2000s. Like earlier projects at PARC and Bellcore labs, The Microsoft Virtual Kitchen [97] linked three workplace common areas (in this case, kitchens) using media connections. A projected image in each kitchen showed feeds from two other kitchens alongside the view from the local kitchen and a television channel designed to attract viewer attention. In response to privacy concerns, the system provided a mechanism to allow users to temporarily disable the camera and microphone. Telemurals [103] provided a more abstract link between two spaces in which video feeds from both connected sites were transformed to provide a single projected view. The Telemurals projection altered to reveal more detail when it detected interactions between the sites.

Media links also remain an area of interest for modern artists; in 2008 Paul St George created the "telectroscope," an outdoor interactive video link between London and New York [7].

2.3   MID- TO LATE 1990s: AMBIENT AND WEARABLE DISPLAYS

2.3.1   AMBIENT DISPLAYS FOR CALM COMPUTING

Inspired by Mark Weiser's vision for ubiquitous, invisible computing [196], a series of projects in the mid- to late 1990s explored methods of ambiently augmenting the environment with information displays.

The Flexible Ubiquitous Monitor Project (FLUMP) [62] used traditional LCDs with the aim of providing a "heterogenous, ubiquitous multimedia information system . . . enabling useful information to follow people around the [departmental] building" [62]. The initial FLUMP deployment consisted of a single "FLUMP station" (see Figure 2.1), i.e., a wall-mounted CRT display with an associated hidden computer [62], and a second station was added later.

To allow content to follow users as they passed the displays, FLUMP used an infrared badge and scanner system—Olivetti's Active Badges [195]. If a FLUMP station detected a user's Active

FIGURE 2.1: The FLUMP deployment—an early adaptive sign.

Badge, it would show that user's personalized homepage. Homepages could include a variety of content items including unread email messages for the user, upcoming appointments, a cartoon and the opening status of a local coffee bar. When no registered users were detected in the vicinity of a display, the FLUMP station would carousel through a sequence of default webpages.

An alternative take on Mark Weiser's vision for ubiquitous or "calm" computing [197] was explored in a number of projects that focused on methods of peripherally displaying information through non-traditional hardware. Natalie Jeremijenko's Dangling String [197] (also known as "Live Wire") was comprised of an eight-foot piece of plastic spaghetti (string) that hung from a stepper motor connected to a nearby ethernet cable. As data was transmitted over the network, the electrical signals caused the motor to turn, resulting in movement of the string and yielding a peripheral indication of the level of traffic. During high-traffic periods the string would whirl round "madly," accompanied by an audible noise from the motor, while in quiet periods only a small twitching movement would be visible.

Bohlen and Mateas's Office Plant #1 [20] took the form of a robotic structure intended to be reminiscent of a small plant. The plant was comprised of a small spherical bulb mounted on a stem and surrounded by wire fronds; a speaker was concealed within the bulb. The plant responded to incoming email by altering the plant's physical posture and could also emit a variety of background noises. The idea of ambient plant displays was explored further in a collection of works in the mid-2000s: the LaughingLily [6] provided an artificial plant mechanized to reflect the types of conversation occurring in a meeting room (silence, productive conversation, argument), while the Living Plant Display [84], Spore 1.1 [53] and PlantDisplay [110] manipulated the environmental conditions of real plants such that their health and growth reflected some additional data sources (e.g., watering the plant to reflect an increase in share prices [53] or denying the plant light to reflect a lack of social interaction between the plant owner and another individual [110] or directing light, and therefore plant growth, to reflect trash/recycling disposal [84]).

There are numerous other examples of ambient displays. For example, the Information Percolator [80] used a series of 32 transparent tubes filled with water in order to create a scrolling display of approximately 32 x 25 bubble pixels. Each tube was fitted with two aquarium pumps that enabled air to be released up the tubes in a precise manner in order to create a bubble to travel through the water. As the bubbles rose up the display it created a natural scrolling effect. The display could be used to display short pieces of text, a simple representation of activity in front of the display or in a nearby corridor, and to gain attention through audio and visual patterns in order to notify those in the space of an upcoming event or of the passage of time.

Rodenstein experimented with the use of Privacy Film–covered windows as a surface for projected images and animation to allow peripheral display of short-term weather forecast data [161]. Their use of windows in this way was intended to complement people's natural instinct to look out for information-gathering and aesthetic purposes. Like Rodenstein's windows, later work on the FogScreen [152, 153, 154, 155, 156, 157] used projection to provide a mechanism for traditional image and animation display on an unconventional medium. A variety of tracking mechanisms were trialed to enable interactivity with the FogScreen [153, 156, 157], and a range of applications suggested including performing arts [154] and advertising [152].

More recently, Breakaway [96] and Clouds [78, 163] designed ambient information displays as a mechanism for changing viewer behavior. Breakaway [96] was a small desk sculpture that moved into a slouching pose to reflect the inactivity of a desk occupant; once the worker took some time away from their desk, the sculpture would return to an upright position. The Clouds installation at the Open University [78, 163] was part of a larger project designed to encourage use of the stairs in preference to the elevator when traveling within a particular building. Twenty-four colored spheres hung from the ceiling: half were orange and represented elevator use, and the remaining half were gray and represented use of the stairs. Each set of twelve spheres could be moved closer to the ceiling or floor to reflect the changing use of stairs and elevator; vertical distance between the Clouds would indicate the difference between the number of people taking the stairs versus those taking the elevator. The installation was supplemented by an array of plasma screens that gave a more detailed representation of stair/elevator usage for the current or previous working week.

2.3.2   WEARABLE PERVASIVE DISPLAYS

While the vast majority of early public displays were static installations, researchers have also explored mobile displays. Perhaps the earliest work on movable displays came in the form of small interfaces that could be worn on the body and displayed information to others. ThinkingTags [24, 25, 27] (part of the GroupWear project) extended conventional name tags by augmenting them with five colored LEDs that allowed a viewer to see the degree to which they shared opinions with the wearer. Infrared communication allowed two badges to exchange answers to five questions—for each answer the two badge owners had in common a green LED would be lit; for each question they answered differently a red LED was lit. These small displays were intended to spark conversations between attendees at events.

Further work in the GroupWear project led to the development of Meme Tags [24, 26]. Like ThinkingTags, Meme Tags took the form of a wearable name tag. Each tag included an LCD screen capable of showing up to 32 characters of text at any time. Infrared communication was again used to support data exchange between the tags—when badges came into range each would display a meme for the other using the name of the viewer, e.g.,

```
Fresh meme for Bob:
Computing should be about
insight, not numbers                                    (Borovoy et al. 1998 [26])
```

Like many modern systems, Meme Tags also provided support for information take-away—upon encountering a meme on another's tag, viewers could transfer that meme to their own to allow further sharing.

Ljungstrand et al. created a wearable display platform, "WearBoy" [116], by adapting the components of a Nintendo GameBoy. Rearranging the components of the gaming device allowed them to create a small wearable device just larger than the 2.6" color screen and weighing less than ninety grams. While much wearable computing work has focused on providing computing for the wearer (e.g., by augmenting reality with an additional information feed) [61, 77, 200], the WearBoy platform formed the basis of two wearable display devices designed to be visible to those other than the wearer. ActiveJewel [116] provided a mechanism for users to express themselves with digital jewelery—the WearBoy device formed a digital brooch that supported moving, repeating and non-repeating patterns as a mechanism for drawing attention and decorating the wearer with a visual representation of their personal values. By contrast, the BubbleBadge [56, 116] was intended to be open to content from the viewer and the environment as well as the badge wearer; scenarios focused on the display of information in the form of relatively short chunks of text. The badge was intended to be worn as a brooch, close to the face, with the intention of providing information to the viewer and promote face-to-face interactions.

Research in the wearable computing space continues but is predominantly focused on providing personal computing services and displays to the wearer (e.g., Google Glass). However, advances in smart textiles have been demonstrated to provide display capabilities [9, 46, 71, 158], and these may provide an interesting technology for future wearable displays.

2.4 EARLY 2000s: PERVASIVE DISPLAYS IN THE WORKPLACE

2.4.1 DOOR DISPLAYS

Throughout the early 2000s, a number of projects began to explore the use of digital displays as doorplates for offices, meeting rooms and shared spaces. Perhaps the first such work, Palplates [119], was deployed at FX Palo Alto Lab and consisted of a set of touch-screen terminals whose interfaces were keyed to their specific location within the lab and were intended to support common tasks. Prototypes by Nguyen et al. [139] initially used a small character LCD and then later a small touch-sensitive color screen to support location specific messages and information (e.g., the last known location of a particular office inhabitant); their "Dynamic Door Displays" would also allow a viewer to leave behind a message for the sign owner.

Two generations of Hermes door displays were developed and deployed at Lancaster University. Hermes I [37, 38, 64] was deployed outside ten offices for 27 months between 2001 and 2004. HP Jornada PDA devices were mounted outside offices within the computing department and connected to a central server using 802.11 networking. Office owners could use their displays to leave a message for passersby (with additional specific messages for particular individuals as identified through Java iButton authentication): messages could be entered on the devices themselves, through a web interface, via SMS or MMS, using tangible buttons or via email. Visitors could leave messages at the device using a stylus and touch screen—their messages could then be retrieved by the owner locally at the device, or remotely using a web interface or email client.

Following relocation of the department to a new building, the Hermes II [39] system featured 40 displays with wider screens than the original deployment. In addition to the original Hermes I features, the devices supported video and audio messages and were designed to support displays with multiple owners in shared offices. A photo display was also introduced.

The Intelligent Mobile Messaging System (IMMS) [13] was deployed outside staff offices within the Computer Science department at the University of Birmingham. The system used handheld devices fitted to office doors to act as information and messaging terminals for students who were hoping to interact with the office occupants. Office owners could update the message on their device using SMS or a web interface, and messages left by students could be received through the same mechanisms (depending on the urgency of the message and the device owners' preference).

Unlike the previous systems in this section, RoomWizard [143] was designed for deployment not in personal offices but shared meeting rooms. Each RoomWizard display consisted of an

FIGURE 2.2: The Hermes digital doorplates.

eight-inch color touch screen mounted outside a bookable meeting room; each display also had a pair of light strips along the side of the casing. The interface would display the timetable of bookings for the day, a number of lines of text describing the room's current status (e.g., "booked for Person X") and also a colored-light indication of the room's availability (green lights indicated availability and red that the room was currently unavailable). Advance bookings could be made at through a web interface (each display also acted as a web server), but ad hoc bookings could be made in situ if the room was available. A deployment of five RoomWizard displays was trialed in two buildings of a large UK company. Such systems are, of course, common in many office buildings now and form one of the application domains we shall explore in Chapter 3.

2.4.2 WORKPLACE AWARENESS

In parallel with projects that explored the use of small public displays as digital door plates, the early 2000s also featured a significant body of work exploring the use of digital displays for improving awareness and developing a sense of community within the workplace. For example, the Learning Communities Newspaper [85] took the form of a web-based application projected in a shared space used by members of the Learning Communities group at Apple. News stories were submitted by group members, via email, to inform other members and guests about their project work and events.

Greenberg described a prototype system, Dynamic Photos [73], in which a group photograph hung in a space could be used to promote awareness of the availability of its subjects and to support transition from awareness to conversation by providing a live video connection. Greenberg used the elastic presentation system [34] to distort a digital image of the group such that the size of a person's face indicated their availability (as detected through sensor data).

The Aware Community Portal [167] at MIT Media Lab consisted of a projected display with an associated camera and server used to display items of relevance to researchers within the laboratory. The display showed live news and weather feeds, an hourly cartoon strip and a periodic clock update; a feed from the camera was also shown. Unlike many similar displays within research workplaces, the Aware Community Portal altered its behavior in response to viewer engagement: as a user walked past the display, articles would cycle through in sequence; if a user stopped to look at the display, the cycle would pause and more detailed coverage of the current article would be shown. In order to promote awareness of others' behavior within the space, the display would show captured images of others looking at the articles on a timeline alongside the articles themselves and would also show a general overview of movement within the space.

The CommunityWall [72, 178] displays deployed at the Xerox Research Center used touch-sensitive SMART Boards [177] to display items submitted by users. Items could be contributed by email, through a web bookmarking system, by scanning a paper FlowPort [203] form, through a Palm application and through KnowledgePump [72], a web-based bookmarking and recommender system. The display would show 10–15 items at a time selected based on a set of rules that assigned each item a priority. The touch interface on the display also allowed viewers to expand, email, print, rate or comment on each item. Evaluation of the deployment of two displays showed that the system had approximately 20 regular users and that at least 50% of articles were interesting enough to users to have received some viewer interaction [72].

The MessyBoard was originally designed with the goal of improving memory and awareness [59], but later development focused on improving communication between coworkers [58]. The MessyBoard was a shared bulletin board space that could be modified via the Web and viewed as a screensaver or projected display. Trials with a number of different groups suggested that use varied greatly and that projection of the display in a public space increased interaction with the board. The MessyBoard was used both for work-related purposes (e.g., arranging meetings or collaborating on projects) and for play (e.g., collaborative games).

The Plasma Poster Network [42] consisted of three touch-enabled plasma displays. The displays were located in the FX Palo Alto Laboratory (one in the kitchen, one in a foyer and one in a hallway) and were oriented in portrait format to reflect the typical layout of traditional paper posters. Content was generated from the laboratory's intranet pages and from items submitted by users via email and the Web. A "PosterShow" interface cycled through content items, and touch-screen buttons were provided for navigating between content items, printing, forwarding and responding to items. The Plasma Poster Network was evaluated through examination of the data collected through ten months' use and through interviews and email surveys. Over the ten months, 859 items were submitted, and users commented that they tended to submit items they thought would be of peripheral interest to others. The authors commented that their observations indicated that the displays had increased social interaction between members of the lab.

Huang et al.'s Semi-Public Display [87] was deployed in the Everyday Computing Lab at the Georgia Institute of Technology with the intention of promoting collaboration and coordination of a small co-located group. The system used a touch-enabled SMART Board [177] and provided four applications: a "collaboration space" for asynchronous brainstorming and discussion around a topic; a set of reminders; an "active portrait" that provides a graphical representation of group activity over time; and an "attendance panel" that provides an abstract visualization of planned attendance at upcoming events. After two weeks of deployment, an initial evaluation suggested that the attendance panel and reminders were regarded positively and were useful in helping maintain awareness within the group while the collaboration board and active portrait were less useful, perhaps because of technical problems.

Moving away from the academic setting, Bardram et al.'s AwareMedia system [11] was designed to raise awareness and support messaging within the surgical ward of a hospital. The deployment consisted of a total of ten clients (mostly large touch-screen displays, some with associated cameras) and Bluetooth-based location tracking. The screens were required to be very information-heavy, with support for social awareness (awareness of a person), spatial awareness (awareness of a place) and temporal awareness (awareness of past, present and future) as well as providing a messaging service between locations. Evaluation interviews three months into the deployment suggested that the displays were useful for asynchronous communication and that the awareness information was useful for informing behavior. Wilson et al. [201] also explored use of a large display in a medical setting. They focused on handover practices between shifts and ran a simple two-week probe in which a digital photograph of existing paper records was projected onto the wall of the handover room.

2.5   EARLY TO MID-2000s: PROMOTING SOCIAL INTERACTION AND COMMUNITY

Following on from themes introduced by early wearable displays a number of display projects in the early to mid-2000s had the specific aim of promoting social interaction. The GroupCast system [122, 123, 124] consisted of a single peripheral display deployed in a common area of the Accenture Technology Labs. Unlike other workplace displays, its aim was not to promote awareness within the department nor was it intended for active engagement. Instead the display aimed to show items of interest to at least one passerby (as identified by an infrared badge system) in the hope of sparking conversations between colleagues.

The Interactive Wall Map [122] consisted of a large wall map (approximately 4m x 2.5m) augmented with three pairs of flat-panel touchscreen monitors placed within different geographic regions and 24 buttons placed over cities of potential interest. Location-related information was displayed in response to user presence or could be explicitly requested by using the button switches. The map was intended to elicit conversation and stories around place and travel.

Brignall and Rogers's shared display, the Opinionizer [31], was designed for informal gatherings and allowed users to add their views and opinions for others to observe and comment on. The system was trialed in two scenarios: a book launch party (two hours, approximately 300 attendees) and a welcome party for new postgraduate students within one school of a university (2.5 hour deployment, approximately 150 attendees). The Opinionizer consisted of a large projection controlled by a laptop, which was also used for entry of the opinions/comments. Interaction with the Opinionizer increased throughout the trials, and many participants were positive about the display as a mechanism for supporting social interaction.

The Dynamo interactive surface [30, 94] consisted of one or more displays, tiled either horizontally or vertically. The surfaces were designed to promote collaboration and could be used as a communal resource with private areas "carved off" for individual or shared use. Interaction with Dynamo could be achieved through a mobile device (e.g., laptop, PDA) or through an "interaction point" consisting of a set of USB slots, a wireless keyboard and wireless mouse. The surface supported a wide range of media types that could be displayed or exchanged. A two-week deployment in a high school common room showed that the system was an effective means of sharing media both by making the files available and also by promoting performance-style interactions [30]. The students quickly appropriated the system for their own purposes.

Systems such as Sparks [41] and Ticket2Talk [125] were developed to encourage conversations at conferences. For example, Ticket2Talk was deployed at an academic conference. Each participating user was equipped with an RFID tag and registered by submitting an image and caption representing an interest or topic that they would be happy to talk with others about. A single Ticket2Talk display was deployed behind one of three tables used for serving drinks and snacks during breaks; as tags were detected at the display they were added to a queue of nearby participants. The display cycled through the queue displaying each person's conversation starters alongside their name and photograph; each one was displayed for five seconds. Evaluation suggested that the display did result in some additional social interactions but that overall respondents did not consider that Ticket2Talk made a significant positive or negative impact on the conference.

Displays continue to be used as a mechanism for promoting social interaction. Farnham et al. [57] deployed their CoCollage display in a community-oriented cafe to support awareness and face-to-face interactions. CoCollage provided tools for the creation of online profiles which allowed the sharing of media and provided a mechanism for online conversations. Cafe community members could then share items using their profiles which were visualized as a continuously updated collage on the screen in the cafe—items from physically present users were prioritized over those from other community members. Within the first month of deployment, 82 users had created accounts. Of these 71% uploaded content for sharing. Between 20% and 30% of users added comments to shared items, commented on profiles and sent instant messages. Users who reported that they would like to make friends were more likely to actively participate in the system.

Another recent deployment, a display-based yearbook system referred to as USIAlumni Faces [164], was found to support the sharing of memories and to stimulate conversation between groups of people meeting around the display. The system used a large screen together with a custom-built input device contracted from a Wii [140] controller and infrared pen. Over two hundred attendees used the deployment of USIAlumni Faces at a university reunion event.

2.6 LATE 2000s: NETWORKED AND LONG-LIVED DISPLAY DEPLOYMENTS

The late 2000s saw an emergence of long-lived deployments in a variety of locations including city centers, rural communities and university campuses.

Commercial display deployments, typically used for advertising purposes, became (and remain) increasingly common in urban spaces. For example, the INFOSCREEN deployments in Germany [185] and Austria [89] first began during the late 1990s, with expansion into Poland in 2009. Similar networks now exist worldwide (e.g., DSN [52] has over a thousand displays across India, and Infoscreen Networks plc [90] owns and operates displays within public transport and urban areas of Kuala Lumpur, Malaysia).

The following paragraphs detail research deployments that may be considered "long lived"—typically those deployed for a number of years. Within an urban context, many now feature as part of the street furniture. For deployments in rural areas and universities, the displays have often become a key tool for the community.

2.6.1 URBAN AND RURAL DEPLOYMENTS

The UBI-hotspots deployment in Oulu, Finland [79, 145], consists of twelve hotspot sites, each including one or two 57" LCD panels (indoor hotspots feature a single outward-facing display, while outdoor hotspots use two back-to-back LCDs to support use from both sides). Each hotspot is also equipped with a loudspeaker, cameras, network access points and an NFC reader. The hotspots were deployed in 2009 and have formed the basis for a number of studies, acting as a "'heavyweight' urban probe" [79, p. 1]. One of the case studies we describe in Chapter 8 was deployed on the Oulu testbed.

A further example of an urban deployment is the CityWall [81, 95, 133, 148, 149] that was deployed in Helsinki in May 2007 and was originally intended to show information during large events (e.g., the Eurovision Song Contest) but subsequently formed the basis for a number of multi-user interaction studies.

In contrast to the above works, the Campus Coffee Display [40, 105] was an indoor deployment and consisted of a single display sited within a a cafe in Newcastle, UK. The display remained in place for over two years and provided information on local cultural events. For researchers, the display acted as a technology probe to explore display engagement behavior.

While the above deployments have focused on urban deployments, there has also been research into the use of displays in rural settings. In 2006, an adaptation of the Hermes photo display (described in Section 2.4.1) was deployed in Wray, a small community in Northern England [40, 190]. The system was modified in response to community feedback, and in 2010 the capacity to submit local advertisements, news and event information was added [39, 188, 189]. The display was positively received within the community, eliciting a number of comments regarding its usefulness for new residents and visitors as well as existing residents [187]. Nnub [159, 160], a similar display system, was deployed in a suburb of Brisbane in 2008 and has since expanded to include a number of displays in different Brisbane communities [142].

2.6.2   UNIVERSITY DEPLOYMENTS

Probably the largest and most well known campus public display deployment is the e-Campus system at Lancaster University [67] that we describe in detail in Section 8.2). However, many universities have experimented with research into digital signage. For example, a combination of "News Displays" and "Reminder Displays" formed the iDisplays [136, 138] deployment at the University of Münster, Germany. Following an initial prototype in 2005, a total of seven iDisplays were deployed in May 2006. Placement of either a News or Reminder display was optimized for the location type—News displays were deployed in entrance areas (highest number of unique viewers), while Reminder displays were located throughout the department building (for repeated viewing throughout the day). Four of the deployed displays were later reused for ReflectiveSigns [135]. Photos, comics, news, short videos and other content items were shown on the displays. Content display was initially displayed randomly, but as face recognition data was collected the displays combined historic view times with the known location of the display and the current time of day to weight the content by expected view time.

The Open University's "Clouds" (see Section 2.3.1, "Follow-The-Lights" (LED navigation display) [78] and "The History" (a tiled plasma display spanning 3m x 3.5m) [163] were also a university-based deployment. The three-display installation was deployed in the late 2000s.

2.7   SUMMARY

In this chapter we have highlighted the enormous amount of prior research that has been conducted in the field of pervasive displays. As with most surveys our coverage is incomplete. We have not, for example, discussed the contribution of the graphics community to work on creating video walls, nor have we described hardware innovations in display technology. However, the chapter does provide coverage of many of the important systems that any researcher in the field would be expected to know.

In closing this chapter, we note that in contrast to many branches of computer science, the field of pervasive displays appears to have attracted multidisciplinary research from the outset—perhaps because of the important roles that both users and content play in this type of research. In the following chapters we build on this survey and discuss specific aspects of pervasive displays such as applications, interaction and systems software.

CHAPTER 3

Applications

3.1 INTRODUCTION

In this chapter we focus on the applications of pervasive displays, both current and future. As we shall see, while existing applications are focused largely on replacing traditional analog signs, it is likely that future applications will provide entirely new functionality that does not have direct counterparts in analog signage.

3.2 CURRENT APPLICATIONS

Excluding research prototypes, most widely deployed pervasive display systems are used for one of four different applications: advertising, information presentation, signage or entertainment.

3.2.1 ADVERTISING

Advertising is a major application domain for all forms of public display and digital signage. It forms part of the general class of advertising called *out-of-home advertising*. According to the Outdoor Advertising Association of America, out-of-home advertising generated annual revenues of $6.388 billion in 2011 [146]. Large networks of screens owned by advertising brokers such as Ströer (http://www.stroeer.com) and ClearChannel (http://www.clearchanneloutdoor.com/products/digital/don/) sell advertising space on signage networks to a wide range of customers. Advertisements form a crucial part of most display networks as they are the principal means by which such networks are funded.

In addition to "standard" advertising solutions of the type that can be seen in many shopping malls and airports, we can identify a number of different types of advertising that take place on digital signage networks:

Point-of-sale displays. Many retailers are moving to digital displays at the point of sale to advertise additional products. For example, companies such as Gas Station TV (http://www.gstv.com) and Pumpflix (http:// www.pumpflix.com/) specialize in digital advertising at gas stations. The displays are embedded in the gas pumps and take advantage of the time that users spend pumping gas to show advertisements. Such advertising has a number of advantages over regular digital signage, including the fact that the audience is of a fairly well-defined demographic and are just about to make a purchase so will have an opportunity to act on any advertising message.

Very large-scale public displays.   Very large-scale public displays have undeniable impact. From the first displays in Times Square in 1928 to recent initiatives such as the BBC Big Screens, these displays attract attention by virtue of their scale.

A number of factors influence the success of advertising on digital signage. Most importantly is, of course, the location of the display. This includes not just making sure that the display is sited in an area with significant foot-fall—but also making sure the display is mounted at the correct height and orientation. Research has shown that this can have a significant impact on the amount of time users spend looking at a display [86]—with eye-level being more effective than the more common overhead style of mounting.

Wherever a display is situated, it must compete for viewers' attention. In 2007 the *New York Times* reported that residents of large cities see up to 5,000 advertisements each day [181] and producers of advertisements for digital signage face an uphill struggle to attract viewers attention. As a result, the challenge facing signage owners is how to create content that attracts attention yet does not promote a backlash from viewers. One approach being explored is to combine advertisements with content that has direct value to viewers. Such an approach has a long history in other forms of advertising; examples include television, where advertisements are interleaved with content, and tourist information, where advertisements are placed around a city map.

In [3] the authors present a table, reproduced below in Figure 3.1, illustrating how such multiplexing of advertisements and content can take place in digital display networks.

As can be seen, content and advertisements can be multiplexed in both time and space, and the switch between content and advertisements can be initiated by either the system or by a user action such as touching or approaching the screen.

A further challenge facing advertisers using digital signage is supporting the "take-away" of information from the display. In print advertising viewers can, for example, be supplied with tear-

Interaction type	Content presentation mode		
	Time multiplexing	**Space multiplexing**	**Integrated**
User-initiated	Full-screen advertising display that switches to a store directory upon being touched	Browsable bus timetable with ads next to the schedule	Interactive ball game with a corporate logo attached to the balls
System-initiated	Looping slideshow of various types of content including ads and information	Ads and information displayed side-by-side on the same screen	City map with embedded restaurant ads

FIGURE 3.1: Multiplexing information and advertisements [3].

FIGURE 3.2: Information boards at York Station, UK. (By Green Lane—own work—CC-BY-SA-3.0-2.5-2.0-1.0, http://creativecommons.org/licenses/by-sa/3.0, via Wikimedia Commons)

off coupons, while the web provides a natural model by offering viewers links to follow for more information or to make a purchase. Advertisers in the digital signage world have experimented with multiple technologies to support information take-away, including QR codes and RFID tags. Recently, digital signage systems that include a vending component have started to emerge. For example, in the UK the supermarket chain Tesco has created a virtual store at Gatwick Airport that enables users to carry out last-minute shopping for home delivery ready for when they return from their trip.

These systems that combine advertising and vending address the problem of take-away but demand significantly more interaction with the user and hence are not suitable in many situations in which users have limited time to engage with the system.

3.2.2 INFORMATION BOARDS

A major use of digital signage is as providing information displays. Such displays are familiar sights in airports and stations across the world (Figure 3.2). Such systems are often highly complex, featuring a very large number of displays; and have almost completely replaced traditional split-flap signage used in these domains (Figure 3.3).

In addition to providing travel information, large-scale public displays are also often used for providing information on the status of systems—for example at IT help desks. In this way, the signs

Flug / Flight		nach / to	über / via	planmäßig / scheduled	erwartet / estimated	Terminal / Terminal	Halle / Hall	Schalter / Counter	Flugs / Gate
						1	B	315-338	B 2
LH	418	WASHINGTON		13 20		1	A	051-278	B
LH	3812	BASEL		13 20		1	A	051-278	B
LH	3524	LINZ		13 20		1	A	051-278	T
LH	6816	STUTTGART HBF.		13 25		1	E	910-914	D
LH	3382	ATHEN		13 25		2	A	051-278	B
XG	1043	BARCELONA		13 25		1	A	051-278	B
LH	636	KUWAIT		13 30		1	A	051-278	A
LH	3218	SANKT PETERSBURG		13 30		1	A	051-278	A
LH	344	BREMEN		13 30		1	A	051-278	A
TP	587	LISSABON		13 30		1	A	051-278	B
LH	1354	STUTTGART		13 35		1	A	051-278	B
LH	3728	ZUERICH		13 40		1	B	315-338	B
LH	406	NEW YORK		13 40		2	D	801-807	
AF	5889	LYON-ANNULLIERT		13 40		1	A	051-278	A
LH	756	MUMBAI		13 40		1	A	051-278	A
LH	4532	LISSABON		13 45		1	A	051-278	A
LH	760	DELHI		13 50		2	D	801-807	D
AF	1619	PARIS CH. DE GAULLE		13 50		1	A	051-278	A
LH	1104	LEIPZIG HALLE		13 50		1	A	051-278	A
LH	3342	ISTANBUL		13 50					

FIGURE 3.3: Traditional split-flap signage. (Snapshot by Hisa Ucda, public domain, via Wikimedia Commons)

are providing a form of public dashboard for the system being monitored. Pervasive displays can also be used to provide information displays such as In/Out boards. Indeed, a very early piece of research into context-aware systems was a context-based In/Out board developed by Anind Dey [50] that allowed users to register their entry to and exit from a building using an iButton. The system then updated a digital In/Out board accordingly.

In the above examples, we have described displays that provide simple information display, but there are, of course, a vast number of public displays that also support interaction. Such displays, often classified as kiosks, typically have a touch screen that enables the user to search for information such as available hotels or tourist information.

3.2.3   SIGNAGE

Traditionally, signs have been used for a wide range of purposes, including identification, navigation and warning. Street signs are a classic example where a set of nationally or internationally agreed on symbols are used to provide information to drivers on issues such as speed limits or road hazards.

The introduction of digital signage as replacements for traditional signs has enabled signage to become dynamic and brought corresponding changes to their usage. For example, the speed

FIGURE 3.4: Variable speed limit sign. (Copyright © David Dixon and licensed for reuse under the Creative Commons License)

limit on a highway used to be posted using a static sign and the speed limit remained fixed over an extended period of time. When temporary changes to the speed limit were required (e.g., as a result of roadwork) corresponding temporary signage was required. The replacement of static signs with dynamic road signs has enabled new ways of managing traffic flows through variable speed limits (see Figure 3.4). In the UK, for example, such signs have become commonplace on major motorways, with the limits being enforced by speed cameras.

Replacing static signage with digital signs has also become extremely popular in office environments. These systems typically provide a range of functions, including automatically updating to reflect the meetings scheduled for a room, and may also provide advanced features such as the ability to leave messages for a room's occupants. Initially conducted as research projects, e.g., by Cheverst et al. [38] (Figure 2.2), such systems are becoming commonplace in offices.

Digital signage has also found applications in navigation. A number of researchers have explored how signage can be used to navigate visitors around a building (e.g., [106]), and recent work has explored how digital signs can be combined with mobile devices to provide individualized directions.

We note that as the cost of digital signs continues to fall we can envisage an ever-increasing number of traditional static signs being replaced by their digital counterparts. For example, fire exit signs might be updated to provide dynamic information on the best escape route in the event of a fire, or future navigation signs might update based on the mobility access requirements of the viewer.

3.2.4   ART AND ENTERTAINMENT

In the previous sections we have focused on important but rather mundane uses of digital signage. However, signage and displays are also used for a wide range of applications in the area of arts and entertainment.

Artists have long seen pervasive displays as an interesting and important artistic medium. For example, one of the first innovative uses of public displays was Galloway and Rabinowitz's (1980)

HOLE IN SPACE
Three days in November, 1980
No one knew it, but the sidewalks of New York City and Los Angeles were about to merge!

The **Broadway** department store, Century City Shopping Center, **Los Angeles**   **Lincoln Center For The Performing Arts**, Avery Fisher Hall, **New York City**

ARTISTS: **Kit Galloway & Sherrie Rabinowitz** (Full credits available upon request)

FIGURE 3.5: Galloway and Rabinowitz's (1980) Hole-In-Space. (From the Sherrie Rabinowitz & Kit Galloway archives)

Hole-In-Space (Figure 3.5), in which they provided a live video link between New York and Los Angeles using public displays situated in empty shop windows.

Today, artists continue to find innovative uses for pervasive displays. Displays form a natural mechanism for showing interesting information visualizations, as illustrated in Figure 3.6, where a display in the Gates Building at CMU shows the Electric Sheep artwork by one of its graduates, Scott Draves.

Displays can also be used for showing short art films. For example, the e-Campus display network at Lancaster University was used to show a number of videos commissioned by the campus's Nuffield Theatre. The artwork, entitled "Finding a Language" [115], was a collaboration between a video artist and a poet and involved the creation of a sequence of seven short videos that explored the notion of space within the context of the university campus. In addition to showing the videos as part of the regular schedule of signage content, visitors to campus could also specifically request the videos using a novel interaction model based on Bluetooth device names [48].

One of the most well-known uses of large public displays for entertainment purposes are the BBC Big Screens. These 25-square-meter displays are situated in a number of UK cities. They are operated as a partnership between the BBC and local councils and show a wide range of content. For example, the BBC has run a number of competitions to enable artists to create content specifically for the screens. The Big Screens are particularly popular during large sporting events because large numbers of people can gather to watch them live at the city locations.

FIGURE 3.6: Information visualization at CMU. (Photograph courtesy of CMU)

FIGURE 3.7: A BBC Big Screen. (Photograph by Editor5807. Reused under the Creative Commons Attribution 3.0 Unported license.)

Despite their popularity, it was announced in 2012 that the BBC would be closing its Big Screens department in 2013 as part of a cost-cutting initiative (http://www.bbc.co.uk/blogs/aboutthebbc/posts/bbc-and-the-big-screens). It is anticipated that local councils will continue to operate the screens.

3.3   FUTURE APPLICATIONS

We have looked at current uses of pervasive display networks but how might future public display networks change our lives? In the following sections we describe a number of scenarios for future display networks. These scenarios help illustrate the variety of uses to which a display network can be put.

3.3.1   EMERGENCY SERVICES

The following scenario was first published in Davies et al. [49].

> Sue realises she has lost her daughter Millie in the local shopping center. She immediately calls the local police who ask her to send them a recent photo from her mobile phone. Within seconds all of the displays in the shopping center are showing a photograph of Millie together with a number to call. As the minutes pass the photographs spread over an ever increasing area—reflecting how far away Millie might be. Five minutes after the first call Millie and her mum are reunited after the child was found by a fellow shopper who recognized her photo from the screen.

There is nothing more frightening for a parent than the thought that they have lost their child, or worse still that their child has been abducted. It is crucial to act quickly, and it is tremendously valuable to be able to enlist the help of others in the hunt for the child. Within a single domain (such as a shopping center) it is usually easy to have an announcement made over a public address system, but without a picture it is difficult for shoppers to recognize the child from just a verbal description. Despite the presence of multiple public displays, with current systems it is almost impossible to instantly circulate a photo so.that everyone knows who they're looking for. In a shopping area where there are multiple display management domains or as the search expands to wider and wider areas, nationally or internationally, we simply lack the technologies to unite display resources to efficiently share this important content and the fine control to target where and when to place it.

Public displays represent a particularly effective form of information dissemination in emergency situations because of their ability to reach the full cross-section of the population in an area. In contrast, national systems such as the UK's Child Rescue Alert or the American AMBER Alert system that focus on television and radio are limited in their effectiveness because they deliver content to viewers at home, not citizens present at the critical location. Broadcasting to mobile devices is also not effective, as it requires explicitly pushing information to users' devices.

Allowing emergency services to appropriate displays in an ad hoc and spontaneous manner to get information to citizens in a timely way could also help in case of natural disasters or terrorist incidents. With a unified display network, all displays within a cordon encircling the area could be targeted. Content can be adapted to reflect the position, size and orientation of the display together with its geographic placement relative to the epicenter of the emergency; for example, displays seen as people travel toward the emergency can warn them to turn away, whereas displays seen while traveling away can give helpful navigation instructions to safe areas.

3.3.2   INFLUENCING BEHAVIOR

The following scenario was first published in Davies et al. [49].

> Jack is six years old and participates in his local walk-to-school programme—an initiative aimed at increasing fitness among school children and addressing childhood obesity. To encourage participation, a simple game has been deployed on the area's public display network. As Jack walks to school he passes a number of displays, each showing a cartoon character that gives him an update on his own progress and that of his friends. By visiting the displays Jack also collects "golden leaves" on his mobile phone—when he has enough of these stickers his school redeems them for a sticker book.

Affecting societal change is extremely challenging for any public agency. Public information is often perceived as unwanted or badly targeted and hence fails to have the desired results. As "push" campaigns are often static content (leaflets) and/or uni-directional (television-based advertisements), there is a technical and financial limit to how well the content can be tailored for a particular audience (e.g., raising awareness of initiatives local to the viewer); it can also be easily missed, discarded or "channel-surfed" away from.

A network of interactive displays offers new possibilities for effectively communicating with a target audience. In the case of Jack and his parents, information is delivered in context as they pass by: it's available because the system detects that they are there and are in the target demographic. The content need only be available on school days and at appropriate school arrival and departure times, helping avoid overexposure. The use of sensing and interactivity with personal devices such as mobile phones, in this case allowing Jack to collect his "golden leaves," allows a new dimension of interactivity not currently available in public information campaigns. This enables new ways of engaging with the populace over social issues such as health and well-being (e.g., fitness, stopping smoking) and for improving awareness and engagement with wider political issues (e.g., canvassing popular opinion). Naturally, location tracking becomes a significant issue in such a scenario, hence the importance of building blocks such as the Tacita system [108].

Of course, using digital signs in emergency scenarios also has associated risks. In many emergency situations power and communications are lost and in these cases a reliance on digital technology may cause serious problems. Similarly, if pervasive displays are used to provide safety-critical information, the need to fully secure these systems against attack becomes even more important. Despite these issues, the scenarios we describe hopefully provide examples of how pervasive displays can be used (in conjunction with more traditional signage) to support emergency response.

3.3.3 LOCAL STRAWBERRY SALE

A small, local fruit shop has a lot of fresh strawberries to sell and it is only a few hours until closing time. The owner knows strawberries are selling well at the supermarket and his price is certainly competitive but he needs to get the word out to the locals that they can buy strawberries for him at a good price. He easily creates a short advertisement and sends it to the new "local shopper incentive" scheme running on public displays around the town. In this scheme shoppers who see his advert can clip a voucher that entitles them to a discount; it also contains the location of his shop. He'll have to pay a small amount to the display owners who show his advert based on every new customer who "takes" a voucher, but he's quite sure selling the strawberries will easily cover this.

Local shops are already deploying displays in their shop windows to attract passing trade. By working together to combine their resources and joining an open public display network, "crowd-sourcing" or a "network effect" could be achieved. Local shopkeepers could reach a mass audience at very low cost. Indeed, the network could support a "micro-value chain" where payments for advertisements are small and easily budgeted for—analogous to the Internet advertising revolution where Google AdWords campaigns offer cheaper, more democratically available and contextually targeted advertisements than traditional broadcast newspaper and TV media.

This scenario also points to the challenge of finding both a viable economic model, as well as a viable trust model, in order to convince display owners to open up their displays to content from potentially unknown sources.

3.3.4 SELF-EXPRESSION AND PERSONALIZATION

The following scenario was first published in Davies et al. [49].

Mike is a fanatical supporter of his favorite rock band and cares deeply that his appearance and accessories match the band. As Mike gets ready to go out to a concert on a Saturday night he chooses his clothes, styles his hair and selects the appropriate images he wants to project onto nearby public displays. As he gets closer to the venue, nearby displays start showing a subtle overlay on their content that hints at the band's logo. Closer to the venue the images become more prominent and Mike is happy to see the images he selected being combined with those of fellow fans to create unique montages that reflect their shared passion for the music.

Many people enjoy making a statement for others to see. Clothes, jewelery and footwear are carefully chosen to show profession, interests, mood or the music or subculture they identify with. Houses, gardens, cars are all accoutrements of social status and objects of conscious or unconscious display. Even graffiti can be seen as a particular form of personalization, in which disaffected youth "tag" locations with their personal messages. Display networks might help channel this creativity and desire for personalization. "Street" applications can emerge that allow for new ways of associating personalization, identity or membership. New types of interactive, situated games can create new activities.

There are many other demographics that can benefit from personalization, and, as with the Internet, the range of applications and potential for innovation is without limit. Singles can hint at their availability in the presence of other like-minded individuals, using the displays as a virtual "aura" to express their moods, intentions or desires. Clubs, societies and bands can find new ways of recruiting members. Those wishing for conversation can find topics of common interest or gain insight about the people around and their interests. Displays can be used in an ambient fashion to intrigue or for the development of new forms of situatuated ambient or interactive public art.

3.3.5 CYBER-FORAGING: TRANSIENT DISPLAY USE TO AUGMENT MOBILE HARDWARE

Dr. Jones is at a restaurant with her family. She is contacted during dinner about a pathology slide that must be interpreted while surgery is in progress. Viewing the slide on her tiny smartphone screen would be useless. Fortunately, a large screen in the lobby (that usually displays advertising) is available for brief use by customers. Walking up to the display, Dr. Jones views the slide at full resolution over the Internet. Using her smartphone for control, she is able to zoom, pan and rotate the slide just as if she were at a microscope in her lab. Privacy-sensitive clinical information about the patient is displayed only on the smartphone. Dr. Jones quickly interprets the slide, telephones the surgeon and returns to dinner.

This scenario, first published in Wolbach et al. [202], presents a very explicit intentional model in which a user cyber-forages for a public display to supplement the capabilities of their mobile device. It illustrates a world in which public displays become part of the general computing and networking infrastructure to be leveraged by passersby, much like a public Wi-Fi access point. One can imagine many other scenarios involving transient use of appropriated displays, perhaps by users many thousands of miles from their home environment. For example, the use of conveniently placed displays to view detailed plans and engineering drawings could aid a visiting expert providing advice in a factory or industrial setting.

3.3.6 PERSONALIZED INFORMATION

Laney is a keen traveller who takes every opportunity to visit new places. She's recently come back from Hong Kong and is working all hours at her three part-time jobs while considering her next big trip: some friends have invited her to join them on a trip to Morocco but she also has an opportunity to go to Russia with an old contact. On a quick break for lunch she can't resist the urge to look in a nearby travel agent's window for inspiration. She glances at the last-minute package deals but nothing appeals to her. In the remainder of the window, a digital display shows flight costs, weather and other local information for a selection of worldwide locations. As she turns her attention to the display she realizes that among the listed locations weather is supplied for Moscow (currently 1°C) and Rabat (currently 18°C)—she'd been researching hostels in these two locations only last night. As she walks away from the travel agent's she thinks some more about her options—a cheap trip to cold Russia, or a more expensive flight to warmer Morocco—it's certainly cold at home right now, perhaps a warmer climate would be nice.

In an earlier scenario, we already described how displays might be used to support self-expression through personalization. In this scenario, we describe a scenario illustrating a different type of personalization, i.e., personalization of the information being presented.

Of course, the mechanisms used to support this scenario may well be similar to those used to support personalization, but the overall effect is quite different. Instead of using the display to broadcast information about ourselves, we are instead effectively issuing a query to the system and viewing the results—it is the difference between posting a web page and posting a web query.

Our example illustrates personalized travel information, but future signage systems may well support a wide range of personalized information, such as menus from a viewer's favorite restaurant or details of upcoming events that are likely to be of interest.

3.4   ANALYSIS

In this chapter we have considered existing and emerging applications for pervasive display networks. Existing applications such as advertising or signage tend to serve as replacements or improvements on traditional static systems. In particular, the ability to show multimedia content and to rapidly change the sign contents for minimal cost enables significant improvements in such applications.

In contrast, the emerging applications we have described are distinguished by having no obvious counterpart in traditional signage systems. They can be seen as entirely new and potentially disruptive uses of the technology. The envisioned applications are typically highly sensitive to their environment and to the viewers in front of them and they blur the boundaries between signage, communications and entertainment.

In the next chapter we describe the models researchers and practitioners use to understand how audiences behave in front of public displays.

CHAPTER 4

Audience Behavior

4.1 OVERVIEW

As displays proliferate in public settings there is an increasing need to understand how viewers (audiences) behave in relation to these displays. Such an understanding helps content producers decide what sort of content engages audiences, interaction designers create appropriate user interfaces and architects and installers understand where to physically situate displays. For example, content on screens may serve to alter the trajectory of a user past the display—and interactive capabilities can lead to people being drawn in by the content, hence stopping, looking for an extended period of time and ultimately starting to interact. This may lead not only to congestion and people colliding but also in more severe cases to people becoming so immersed in the interaction that they forget about their surroundings with potential safety implications. For all these reasons, it is important to understand audience behavior and how it can be controlled and exploited by the content shown on displays.

4.2 AUDIENCE BEHAVIOR MODELS

Public display researches have long tried to understand what makes users engage with displays and have developed a series of models that try to explain audience behavior. These models typically try and capture two distinct aspects of a viewer's relationship with a screen:

Spatial aspects. Spatial models try and describe how audiences behave in terms of viewer's spatial relationships to a display.

Temporal aspects. Temporal models try and describe how audience engagement with a display changes over time.

In addition to considering the temporal and spatial relationships between viewers and displays, some models also endeavor to capture the relationships between multiple viewers—for example when a passerby watches a person interact with a display and then proceeds to engage with the display themselves.

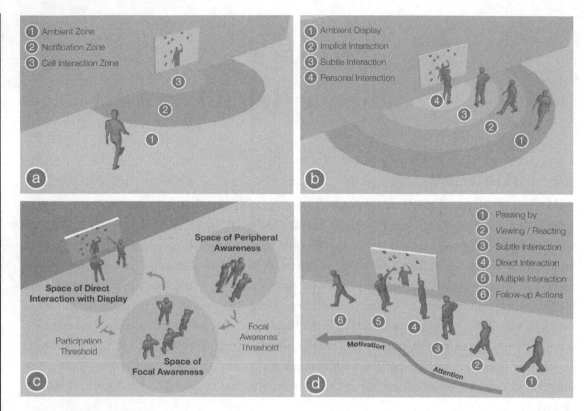

FIGURE 4.1: Interaction models for public displays: (a) Three Zones Model (Streitz et al. [184]); (b) Extended Model (Vogel and Balakrishnan [192]); (c) Public Interaction Flow Model (Brignull and Rogers [31]); (d) The Audience Funnel (Michelis and Müller [131]).

4.2.1   SPATIAL ZONE MODELS

The earliest work that specifically attempted to provide a spatial model of interaction with a digital public display was by Streitz et al. [151, 184]. This was conducted in the content of their work on "Hello.Wall," a wall-sized ambient display that employed what the authors termed Viewports. These Viewports could not only sense users in the vicinity of the display but also enabled interaction with the display based on RFID and WaveLAN. The authors developed a spatial interaction model based on three zones: an *ambient zone*, a *notification zone*, and a *cell interaction zone* (Figure 4.1(a)). This model was mainly used to define the interactions offered, and the kind of information to be shown, on the display.

Vogel et al. [192] refined the original zone model by further dividing the cell interaction zone into *subtle* and *personal* interaction zones and by generalizing the idea of a notification zone into an *implicit interaction zone* (Figure 4.1(b)). As a neutral state, they also defined a so-called *ambient display* to be an anchor point for subsequent interactions. Even though the model is mainly concerned with how people relate to the information presented on the display, there is a noticeable relation to Hall's theory of proxemics that defines four zones of interpersonal distance (public, social, personal, intimate) [74]. Saul Greenberg's group at the University of Calgary has explored the issue of proxemics and public displays in detail; see [194] for an example.

Both of these spatial models are strongly geared toward information presentation and they are mainly suited to modeling single-user interaction. They do not provide mechanisms for modeling how users move back and forth between the different zones.

In addition to modeling relationships between viewers and displays, research has also looked into modeling how public displays can be used to stimulate interaction between people in the vicinity of the display (friends, family member, strangers). In their engagement zones model, Memarovic et al. distinguish between a *passive engagement zone* and an *active engagement zone* [127]. In the passive engagement zone, people observe others interacting with the display (observations) or engage in passive interaction such as briefly reading the content (glimpse interaction) without actively engaging with the environment. In the active engagement zone, people engage more actively with the display—for example through active reading or prolonged, explicit interaction. Once people are actively engaged, social triangulation may occur (e.g., a discussion about the content on the display) and people may start transitioning between the active and passive engagement zones.

The strength of the engagement zone model is that it considers multiple users and allows for modeling back-and-forth transitions between the different phases.

4.2.2 THE PUBLIC INTERACTION FLOW MODEL

The public interaction flow model was developed by Brignull and Rogers [31] in order to describe how viewers engaged with a system called Opinionizer. Studying engagement with the system at two social events enabled the authors to explore how groups socialize around public displays, how they walk up to them and how they change roles, i.e., change from an onlooker to an active participant.

Brignull and Rogers identified three activity spaces. The *peripheral awareness activity* describes situations in which users are primarily socializing or eating/drinking. They are generally aware of the presence of the display, yet are not focused on it. In *focal awareness activities*, the attention of the user shifts toward the display, e.g., people start talking about the (content on the) display and watching the screen. Finally, during *direct interaction activities*, individuals or groups of users start to explicitly interact with the display, e.g., the system allowed users to post content. Identifying and investigating the thresholds that lead to people switching between the activity spaces allowed the authors to develop the conceptual framework of public interaction flow, depicted in (Figure 4.1(c)).

As people are motivated they cross the threshold from peripheral to focal awareness, and when quick and enjoyable interaction techniques are provided users cross further into the participation threshold.

The strength of the framework is that it (i) supports multi-user interaction and (ii) takes people moving between the different activities into consideration. In contrast, it disregards both implicit interaction and explicit interaction from a distance (probably due to the use of keyboard and mouse as input devices).

4.2.3   THE AUDIENCE FUNNEL

Michelis et al. [131, 134] extended the public interaction flow model and developed the audience funnel, which focuses on observable audience behavior (Figure 4.1(d)). The model consists of several interaction phases and specifically attempts to model the probability of users transitioning between phases (the likely "conversion rate"). In the first phase, users are simply *passing by*. As these people notice the display they may start *viewing or reacting* to it, for example by turning their head. For displays that support interaction based on gestures, movement, or presence, the *subtle interaction* phase is when people are trying to understand the effect they have on the display, e.g., through waving their hand. As users engage further with the display, *direct interaction* occurs, and in this phase users often position themselves directly in front of the display. *Multiple interaction* is also possible in situations where multiple displays are available or if viewers chose to come back and interact with a display at a later point in time. Finally, *follow-up* actions may take place including, for example, taking a picture in front of the display or entering the store where the display is deployed.

Between each of these different phases, certain thresholds can be identified that need to be overcome in order to enter the next phase. For example, to overcome the first threshold and transition from "passing by" to "viewing and reacting," the passerby's attention must be captured. To overcome the second threshold and move on to "subtle interaction," the onlooker's curiosity must be piqued. Subsequent thresholds may require that users are supplied with additional motivation.

The strength of this model is that it can be used to describe conversion rates and thus provide a measure of success for public display content or applications. This is particularly interesting to advertisers. However, the model does not consider social interaction between viewers. Furthermore, the model describes audience engagement as a rather linear process in which people are unlikely to move back and forth between the different stages and the model.

4.3   ENGAGING USERS

Modeling user behavior is one thing, but how does the designer of a public display engage users with their system? More specifically, in terms of the models we have discussed, how does a designer move potential viewers through the various phases of engagement? This problem is particularly important for interactive signage systems where success is often measured in terms of interactions. Based on the various audience models described above we can identify three basic issues that must be tackled: attracting the viewer's attention, communicating the potential for interactivity, and motivating further engagement.

4.3.1   UNDERSTANDING ATTENTION

Human–computer interaction techniques often assumes that the user is aware of the computer in the first place. This is not necessarily the case for public displays. In contrast to many other computing technologies, public displays are not owned by their primary users (the audience)—rather, they are installed in public places where they compete for audience attention. There has long been discussion on how much attention ubiquitous computers should attract. On the one hand, it has been argued that if the environment is filled with ubiquitous computers, they should remain calm and slide effortlessly between the center and periphery of attention [197]. On the other hand, it has been argued that they should engage people more actively in what they do [162]. This debate is not well understood but is likely to prove important in the future if public displays are to become as pervasive as we might hope.

Attracting *attention* with public displays and kiosks is not easy [86, 102, 134] and is described as the "first click problem" [102]. Huang et al. observed passersby's attention toward (non-interactive) public displays and show that most displays receive little attention [86]. One solution is to use stimuli to attract attention [86, 134]. However, this is challenging in public space. Moving stimuli attract attention but do not guarantee that the user looks, because there are many objects competing for the attention of the passerby [86]. Another solution suggests using physical objects. For instance Ju and Sirkin [102] show that a physical attract loop (animatronic hand) is twice as effective as a virtual attract loop (virtual hand). While physical objects seem to attract more attention than digital content, they are less flexible, more difficult to update with new content and may well be reliant on significant novelty value. An overview of the role of attention and motivation as requirements for public displays is provided in Müller et al. [134].

4.3.2   MANAGING ATTENTION

There are a number of techniques for attracting and managing attention that are useful in the context of public displays.

Behavioral Urgency

Franconeri and Simons [65] hypothesize that attention is captured by stimuli that indicate the potential need for immediate action. It has been found that the abrupt appearance of new objects [100] and certain types of luminance contrast changes [54] capture attention. In addition, moving (toward the observer) and looming stimuli have been found to capture attention [65]. Since all of these stimulus properties hint at the potential need for immediate action (e.g., an animal approaching), behavioral urgency may be a useful model to predict how much attention a stimulus will attract.

Bayesian Surprise
Itti et al. [92] propose a model of Bayesian surprise for bottom-up visual attention, which measures the difference between posterior and prior beliefs about the world. They implemented a model of low-level visual attention based on Bayesian surprise to predict eye movement traces of subjects watching videos. The model performs better than other models, predicting attention based on high entropy, contrast, novelty or motion.

Honeypot Effect
The Honeypot Effect [31] is described by Brignull and Rogeres in the context of the Opinionizer public display when it was deployed at a party. Whenever a crowd of people gathered around the display, the crowd seemed to attract further attention, drawing more people to the display. Similar effects were observed with the CityWall display [148] and the Magical Mirrors installation [129].

Change Blindness
In contrast to the previous techniques that focus on gaining attention, change blindness is an effect that describes how to reduce the attention-attracting effect of changes in display content. This may be important if we wish to change the content of a display without viewers noticing its change. Research has shown that in certain circumstances people have surprising difficulty in observing seemingly obvious major changes in their visual fields, e.g., road lines changing from solid to dashed or a big wall slowly changing color. Visual effects that cause change blindness include blanking an image, changing perspective, displaying "mud splashes" while changing the image, changing information slowly, changing information during eye blinks or saccades or changing information while occluded (e.g., by another person). Intille [91] proposes using change blindness to minimize the attention a display attracts while updating content.

4.3.3   COMMUNICATING INTERACTIVITY

If a public display is interactive, then the owners and designers have a significant challenge in communicating this fact to potential users. Today, people do not expect public displays to be interactive—an expectation that has been reinforced by the fact that many displays are being used primarily for advertising. If public displays fail to communicate that they are interactive, they may not be used at all. This issue may become even more apparent in the future, as current LCD technology used for public displays may be replaced with technologies that more closely resemble traditional paper. As a consequence, passersby might not even notice that a surface is digital unless the content is constantly moving and/or reacting to the audience.

Interactivity Cues

A number of techniques for communicating interactivity of both public displays and tabletops have been proposed in prior work:

Call-to-action.   Most common is a call-to-action, often a simple text label such as "touch screen to begin." Examples can be found in Ju et al. [102], Kules et al. [109] and Marshall et al. [120].

Attract sequence.   An attract sequence was originally described as a slideshow [109], but a range of different attract sequences are possible. For example, some multi-touch installations use constantly moving objects [83, 148], while many arcade machines use a video that either explains the interaction or shows a user performing the interaction. A similar technique is to use constantly moving content, as described in Peltonen et al. [148].

Analog signage.   Nearby analog signage, either with a simple call-to-action or a more complex set of user instructions, has been used in many deployments, e.g., [109, 120, 148].

Honeypot Effect.   The Honeypot Effect [31] describes the effect of people being attracted by persons already interacting with a device. Brignull et al. observed this effect and divided the actions of people around the display into the phases peripheral attention, focused attention and interacting. Further observations of the Honeypot Effect are reported in [120, 148, 131]. For the CityWall [148], for example, it was observed that people most often notice the wall when someone is interacting with it (in 19% of the cases). In the absence of others interacting with the system, some passersby had difficulty realizing that the display was interactive.

Inviting passersby.   Persons inviting passersby to interact can be either users who have already noticed the interactivity and now motivate their friends [120, 148], or researchers standing next to the device inviting users and explaining the interaction [95]. Students are employed as so-called UbiGuides in Oulu, Finland, in order to encourage people to use the displays [144].

Prior knowledge.   Prior knowledge that a device is interactive can be used if users pass by the same device multiple times, or if they are familiar with the device (e.g., the Microsoft Surface [120]).

User representation.   Showing a representation of the passerby on the screen is a powerful interactivity cue. Shadow or mirror images have been used in the context of artistic installations [107], pointing tasks on large displays [175] and interaction above a tabletop [82]. Michelis and Meckel [130] deployed public displays showing a camera image of what was happening in front of the screen, and Müller et al. showed that when displaying representations of users the degree of abstraction plays an important role in determining levels of engagement—with increasing abstraction the ability of the user to recognize themselves on the screen (and hence their engagement) decreases [137].

Immediate Usability

After people notice interactivity, immediate usability is important. The term "immediate usability" was introduced in the context of Shneiderman's CHI photo kiosk [109]. Shneiderman derives three recommendations that are applicable to pervasive displays:

1. Implement an attract sequence tailored for the audience; clearly indicate how to end the attract sequence and begin using the system (e.g., using a call-to-action such as "touch screen to begin").

2. Support zero-trial learning. Users should be able to use the interface after observing others or using it themselves for a brief period of time (15–60 seconds).

3. Encourage users to immediately interact with the content.

Users who were not immediately successful would often simply abandon the device. Marshall et al. observed that even a delay of a few seconds after touching an interactive tabletop is problematic [120]. Users are likely to give up and think that the device is not interactive or broken.

4.3.4   MOTIVATING FURTHER ENGAGEMENT

Traditional paper-based public displays have served as read-only media (e.g., posters and billboards). With new, interactive displays, the question of how to motivate users to interact with displays arises. Typically, people do not go out in order to look at public displays. Instead, they tend to come across a public display (e.g., while waiting at a bus stop) and become motivated by external factors to look at it.

The entry of interactive displays into public space is part of a greater trend: the spread of computer usage from the workplace into public life. As such, displays will serve a range of functions, from helping users achieve tasks to more speculative forms of interaction. Traditional task-oriented theories simply regard the "how" of an activity and not the "why," they leave questions concerning underlying motivations unanswered [174]. Malone presents a distinction between tools and toys to differentiate systems that have an external goal from those that are used for their own sake. Tools are task oriented. They are designed to achieve goals "that are already present in the external task." Toys either need to provide a specific goal or enable the user to create their own, emergent goal. A tool should be easy to use, while a toy needs to provide a challenge that motivates the user [118].

Displays force us as researchers to think of the "why." In spite of its increasing significance in human–computer interaction, there is still a significant need to advance our understanding of the motivation behind users' activities [117] and very little is known about how the design of public displays will influence what motivates people to interact [1]. Based on the work of Thomas Malone, who investigated motivating principles for traditional human–computer interaction [117], a list of potential motivators for audience engagement has been produced by Michelis [129]:

Challenge and control.   Challenge can be used to help motivate users to interact but must be carefully balanced—too little challenge leads to boredom and too much challenge leads to anxiety. In human–computer interaction, people strive for an optimal level of competency that allows them to master the challenges presented by the application [29].

Curiosity and exploration.   Curiosity appears to be a key characteristic of intrinsically motivating environments. In order to stimulate curiosity and to influence motivation, the interaction should not be designed in a way that is either too complex or too trivial. Interactive elements should be novel and surprising but not incomprehensible. On the basis of their prior experiences, the user should have initial expectations for how the interaction proceeds, but these should only be partially met [117].

Choice.   Choice as a motivating factor is based on the observation that the motivation for a particular behavior appears to increase if, in the process, people can select between alternatives in behavior and effects can be controlled [93].

Fantasy and metaphor.   In general, imaginary settings also appear to have a motivating effect on behavior. In these fantasy settings, the constraints of reality are switched off so that one imagines possessing new abilities. The extent to which interactive environments inspire fantasy determines their attractiveness and generates interest in the reception of the interaction [147].

Collaboration.   In contrast to the motivating factors already discussed, collaboration is based on interaction with other human beings. A condition for its motivating effect is the opportunity that the individual can influence the interaction of other people [29]. Pervasive displays have the potential to play an important role in fostering collaboration in public spaces.

4.4   OBSERVED EXAMPLES OF AUDIENCE BEHAVIOR

To conclude our discussion on audience behavior, we present a series of examples of observed audience behavior.

4.4.1   THE SWEET SPOT

In 2010, Beyer et al. conducted a study investigating differences in audience behavior based on display shape [16]. To do so, they implemented an interactive application that allowed the user to draw flower patterns on the screen by altering their position in front of the display. This application was then deployed on both a planar display as well as on a cylindrical display covering the same floor real estate.

The data analysis revealed a small area in front of the planar display where participants got themselves in a frontal position—the so-called sweet spot. This area was positioned centrally in front of the display, about 1.5 meters away from it. People who decided to approach the display entered this area quickly and stopped with their shoulders parallel to the display, facing the display frontally. From this position, they could see the entire screen from the best perspective, while the entire frame was still in the visual field.

No such sweet spot was discovered for the cylindrical display. As a result, the authors propose that flat displays, because of their sweet spot, may be more suited for waiting situations and longer dwell times, and may support more complex content.

4.4.2   THE HONEYPOT EFFECT

A quite well-known phenomenon occurring around public displays is the so-called Honeypot Effect. It describes the effect of people being attracted by persons already interacting with a device. Brignull et al. [31] observed this effect and divided the people around the display into the phases peripheral attention, focused attention, and interacting (cf. Section 4.2.2). Further observations of the Honeypot Effect are reported in [120, 131, 148].

Müller et al. investigated the Honeypot Effect in detail in the context of the Looking Glass project that enabled viewers to play a simple ball game. They observed many situations in which different groups started to interact with the game. The first group (or person) usually caused the Honeypot Effect. People passing by first observed somebody making unconventional movements while looking into a shop window. They subsequently positioned themselves in a way that allowed them to see and understand the reason for these movements—usually in a location that allowed both the persons interacting as well as the display to be seen; see Figure 4.2. In this figure, a man interacting with the display with expressive gestures attracts considerable attention. The crowd stopping and staring at him and the display partially blocks the way for other passersby. Newcomers seem to be first attracted by the crowd, then follow their gaze, then see the man interacting and follow his gaze, repositioning themselves so they can see both the man and the display. They also seemed to prefer to stand a little bit to the side, so that they are not represented on the screen.

From this finding, Mueller et al. conclude that the Honeypot Effect is a very powerful cue to attract attention and communicate interactivity: displays that manage to attract an initial set of viewers are likely to be able to grow this audience. The Honeypot Effect even works after multiple days, as people who have seen somebody interacting previously may also try the interaction in the future. It seems that for maximum attention displays should be designed to have someone visibly interacting with them as often as possible. Since the audience repositions themselves such that they

FIGURE 4.2: Audience repositioning.

FIGURE 4.3: Drag-to-screen.

can see both the user and the display, the environment also needs to be designed to support this, e.g., through being visible from a wide angle or by providing considerable space in front of the display.

4.4.3 THE LANDING EFFECT

Another phenomenon unveiled during the Looking Glass project was the Landing Effect. When investigating the moment when people start to interact, Mueller et al. found that people often stop late, i.e., as they have already passed the display. Figure 4.3 shows this effect with a couple, and Figure 4.4 with a group. In Figure 4.4, a group of young people are passing the display. One of the last members of the group looks at the display but keeps on walking with the group for several meters. This group member then suddenly turns around and walks back, followed by a second person. These group members then start to interact with the display and are soon joined by other group members.

The Landing Effect can lead to conflicts when one person in a group notices the opportunity for interaction. It was observed that if the people leading a group suddenly stop and turn around, the

FIGURE 4.4: Drag-to-screen, group scenario.

people following would sometimes bump into them. More often, the whole group stopped rather than walking on. However, when a person at the tail-end of a group noticed interactivity, the leaders would usually walk on for some time before they noticed that somebody had stopped. This situation created a tension in groups as to whether people who had continued walking returned to the display or whether the person interacting would abandon the display and join the group. In some cases the group simply walked on after waiting some time, causing the person interacting to continue playing for only a short moment and then hurrying (sometimes even running) to rejoin the group.

The Landing Effect appears to be a sufficiently strong phenomenon that it can be argued that pervasive displays should be designed with this feature in mind. Specifically, displays should be placed so that when people decide to interact they are still in front of the display and do not have to walk back. Optimally, display deployments should be designed to avoid fragmenting groups by providing support for interaction regardless of whether the initial engagement was by viewers at the front or the back of the group. This could be achieved by, for example, designing very wide displays (several meters), or more practically, a series of displays along the same trajectory. Alternatively, displays could be placed in a way that users walk directly toward them.

4.5   SUMMARY

In this chapter we have provided an introduction to understanding audience behavior around public displays. Research has shown that audience behavior around public displays is very complex and depends on a wide range of factors, including the display size, the display shape, the content and the interaction techniques used. Perhaps more importantly, we need to recognize that audience behavior around public displays is likely to change significantly over time as viewers become more accustomed to the new types of functionality that future displays will be provide. As a result, the models of behavior that we have described may well stay current for some years but the validity of individual observations may be short-lived. Nevertheless, the strength of effects such as the Honeypot Effect and the Landing Effect suggest that these need to be taken into account, and ideally capitalized on, in considering the design and deployment of display systems.

CHAPTER 5

Interaction Techniques

5.1 OVERVIEW

While most existing digital signage is not interactive, there is an increasing interest in supporting interaction between users and displays. By providing support for user interaction, display owners and content creators can enable a wide range of new features, such as:

Navigation. Support can be provided for users to navigate content on the display. This can range from the simple equivalent of changing a channel to reproducing web navigation.

Expression of interest. Allowing users to express interest in the content being shown allows display owners to tailor the content they show.

Content take-away. Interaction can be used to support content take-away—either to a mobile device or to a URL or an email address.

Content upload. Interaction can also support upload of content—important in applications such as notice boards [2].

In the past some public displays were augmented with physical hardware to support interaction such as buttons, joysticks or even keyboards. However, this type of hardware has largely been replaced with more modern technologies such as touch screens. In this chapter we consider the three main types of interaction with displays, i.e., touch, gesture and mobile device–based interaction.

5.2 TOUCH

5.2.1 ISSUES IN SUPPORTING TOUCH

Touch has been used in the context of public displays for many years. Indeed, work on touch screens began in the 1960s and has continued apace since then. Before we begin to examine the touch options available, it is worth noting that touch in the context of public displays raises a number of fundamental challenges.

First, a key part of having a touch-driven interface is that users must be able to physically reach the screen. For many public display installations this is not possible. Traditionally, public displays have been mounted specifically so that they are out of reach—typically by fixing them above head

FIGURE 5.1: An example deployment of a very large public display with touch capabilities.

height. The rationale for mounting screens high up is usually two fold—first it helps ensure that the display is not easily blocked by people and hence is visible from some distance away (though current thinking suggests that screens mounted high up are not good at attracting attention [86]), and second, it helps prevent damage to the screen (either accidental or intentional). If we wish users to be able to interact with a touch screen, then we must rethink where we mount screens.

Situating a display where it is convenient to touch then raises issues of screen protection, screen visibility and screen maintenance. In terms of protection, screens mounted within reach must be (i) secured to prevent theft, (ii) configured to prevent access to controls and (iii) be resistant to everyday knocks that may occur. Screen visibility is an issue for two reasons: first, because screens mounted at eye level or below are less visible from a distance, and second, because viewers interacting with a touch screen are extremely likely to occlude the display for other viewers. Finally, touch screens may also require more maintenance than traditional displays as their surfaces must be cleaned to maintain an attractive appearance that encourages users to interact.

Touch screens also face challenges associated with audience expectations. As discussed in Chapter 4, there are presently no commonly agreed ways of communicating interactivity to the user and most users assume that screens are not interactive. Crucial to the success of touch screens is their ability to communicate to users the very fact that they support interaction. An important second issue relating to audience expectation is the fact that user's expectations of touch capabilities are now set very high—as a direct result of the success of multi-touch phones. While multi-touch is common in smartphones, its use is much less common in the context of public displays. This is a result of a number of factors including both the touch-screen technology used and the support for multi-touch by the operating system being used to control the screen. For example, many displays are driven by low-cost embedded devices that do not have OS support for multi-touch. However, modern displays such as the Philips Giant Tablet and BDT Multi-Touch Series can combine with Windows 7 and Windows 8 to provide support for multi-touch. Interestingly, as pointed out in [5], many users are now expecting screens to be multi-touch and tend to try and use gestures that they know work on smartphones (e.g., pinches and flicks).

Finally, we note that their size and situation makes pervasive displays more liable to have multiple users interacting with them simultaneously than touch screens on traditional phones. In this regard public displays have much more in common with tabletop/surface computing as typified by DiamondTouch [51] than phones.

5.2.2 TOUCH TECHNOLOGIES

A wide range of touch-screen technologies have been developed. Most early public display touch screens used resistive touch screens. In this type of system the screen is covered with two layers of resistive materials, separated by an air gap. When the screen is touched the two layers are brought into contact and the location of the touch can be calculated. Resistive touch screens are cheap to manufacture, can be developed as overlays for existing screens and, perhaps most importantly, can register touch from any object. While resistive screens can be very accurate, they are not as responsive to touch as screens based on capacitive sensing.

Touch screens based on capacitive sensing use an entirely different mechanism for determining the location of a touch. A glass plate that covers the front of the display is coated with a transparent conductor. Contact with another electrical conductor (such as a finger) creates a change in capacitance that can be used to calculate the position of the touch. Compared to resistive technologies, capacitive screens are relatively inaccurate and must be touched by something conductive (i.e., they don't work with a standard stylus or other object). However, they do not require pressure and hence tend to provide a better user experience—particularly for gesture-based interaction.

Other systems available include IR-based touch screens and optical touch screens. In both cases sensors around the edge of the screen detect the presence of a finger or stylus and calculate the position of the touch.

Recent years have also seen an explosion in the popularity of large-scale interactive surfaces. Early systems such as the MERL DiamondTouch table [51] offered multi-user touch support and, uniquely, enabled identification of the user making the touches. However, the system required capacitive coupling between the table and the user (this typically took place via the user's chair that was electrically coupled to the interactive surface). More recently, systems such as the Microsoft PixelSense [132] have shown the potential for large-scale multi-user, multi-touch interaction.

For an excellent overview of some of the issues raised by multi-touch technology the reader is referred to Bill Buxton's well known web summary, available at http://www.billbuxton.com/ multitouchOverview.html. Gesture technologies such as the Microsoft Kinect are surveyed in Section 5.3.2.

5.2.3   EXAMPLE APPLICATIONS

Touch screens have been widely used in public displays for many years and a wide range of applications have been developed. A very common use case is in public-access information terminals of the type often found in train stations and tourist information centers. A recent example are the interactive touch screens launched in Boston that provide a range of local travel and city information for users of the subway system. Initial trials of the system were reported as being very successful, and in 2013 it was announced that the system would be expanded with a further 77 touch screens [8].

5.3   MID-AIR GESTURES

In many cases, touch is not applicable as an interaction method for public displays—for example on very large displays or for displays at a distance that cannot easily be reached by the passerby. Furthermore, people are also often reluctant to use touch screens for hygiene reasons, particularly in public spaces. In such cases, gesture-based interaction provides a suitable alternative that can at the same time also serve as a catalyst for performative interaction [191] or as a cue for the interactive capabilities of a display. However, selection and text entry based on gestures is often difficult, and at present there is no commonly accepted gesture set for public displays.

Gestures—"a motion of the body that contains information" [111]—have been extensively researched in the context of pen- and touch-based interfaces, but research on using mid-air gestures in front of displays is relatively scarce. The following section provides a brief introduction to the challenges of supporting gestures on public displays and an overview of technologies that have been used for detecting gestures on public displays.

5.3.1   ISSUES IN SUPPORTING GESTURES

Overview

Based on their function, gestures can be categorized into epistemic (tactile or haptic exploration of the environment), ergotic (manipulating objects) and semiotic gestures (communicating meaningful information) [32]. The latter two are of particular interest for gesture-based interaction with displays: ergotic gestures can be used to directly manipulate virtual objects shown on the screen (e.g., punching a virtual ball), while semiotic gestures are useful when it comes to executing commands; for example through pointing at a button shown on the screen.

There are no standard or even commonly accepted gesture sets for interacting with public displays, so researchers and developers often need to create their own gesture sets. The following points should be borne in mind when considering the development of gesture sets. First, new gesture sets should ideally draw upon existing operations that users are familiar with. Popular examples include the pan or pinch gesture commonly used on smartphones. Second, gestures belonging to a set should be *coherent* [199], thus making it easier for the viewer to understand which gestures are supported by a system. Gestures also need to be easy to recognize and easy to teach—two topics that we explore in more detail below.

Recognizing Gestures

In order for gestures to be recognized, the beginning of the gesture (registration), its continuation and the end of the gestures (termination) need to be clearly delimited [12, 66, 199]. While these phases can be easily recognized on a touch-enabled surface by sensing when a finger touches the screen, is swiped over it and is finally released, the lack of an obvious delimiter for mid-air gestures makes the recognition of gestures a challenging task. Prior work by Wigdor suggests *multimodality, reserved actions* and *clutching* as possible solutions [199]. Using multiple modalities would, for example, include combining touch and speech—e.g., saying "put that . . . " while pointing with the finger toward an object on the screen and then pointing at the target destination while saying ". . . there!" [21]. However, providing means to discover these additional modalities introduces new challenges, and certain modalities may be inappropriate for the use in public environments (e.g., speech). Reserved actions, such as drawing a certain form in the air, could be interpreted as a command. For example, drawing a question mark in the air could bring up a help menu on the screen. Finally, clutching is the idea that users need to explicitly signal to the system that it needs to engage the gesture recognizer (the term comes from the parallel of engaging the clutch in a car).

Teaching Gestures

A major challenge in gesture systems is how to inform viewers of the display of the gestures they can use to interact with the display. As many gestures are not intuitive, visual clues should be provided to help users to discover possible gestures [141]. As Walter et al. suggest in [193], cues could be presented using spatial division/multiplexing (splitting the screen into one area with content and one where the gesture is explained), temporal multiplexing (the regular content is temporally interrupted with the gesture cue), or the cue could be integrated with the content [193]. Walter et al. found spatial multiplexing to be most suitable for public displays, because it does not interrupt the content and interruptions are known to increase the tendency for users to leave the screen.

Prior work on teaching gestures has suggested a number of techniques for mouse (e.g., [28]) and touch interfaces (e.g., [66]). For example, GestureBar shows a video of how to execute the command via a mouse gesture as the user clicks an icon in the toolbar. ShadowGuides displays various hand poses for registering a gesture as users touch the screen beyond a dwell time of 1 second. These approaches are difficult to apply to public displays for three reasons. First, they require prior knowledge about the modality. Second, these approaches were designed for goal-driven applications where users are aware of the available commands. In contrast, the use of public displays is often not goal driven but occurs out of curiosity or in a playful manner. Third, people in front of public displays are usually neither aware of its interactive capabilities nor how they can interact or whether gesture-based interaction is supported.

Concrete solutions suitable for use in public displays have been suggested in the context of the Public Ambient Displays, LightGuide and StrikeAPose projects. Vogel et al. show self-revealing help in the form of looping video sequences on the screen that demonstrate the interaction [192]. LightGuide is a system that projects guidance hints directly onto the user's hands [179]. StrikeAPose is an interactive public display game that introduces a full-body pinch gesture that requires users to touch their hip with their arm [193]. This gesture is particularly easy to be recognized by the system due to the large enclosed area and the frontal orientation of the viewer.

5.3.2 GESTURE TECHNOLOGIES

A number of different technologies have been explored for gesture-based interaction. Most popular are device- and camera-based techniques.

Device-based techniques rely upon sensors worn on the body, such as accelerometers or positioning sensors. For example, Nintendo's Wii controller or BeeCon's BlueWand [68] use accelerometers to detect movements. Vogel et al. use a Vicon tracking system with markers attached to the body and fingers of the user to enable hand gestures [192].

Camera-based techniques are often favored in public spaces as the user does not rely upon additional gear. These techniques require one or more cameras to capture the scene in front of the display and use image analysis techniques to extract information on body posture or position of the user. Currently, the most popular image analysis technique for gestures is skeleton detection, where information on body, hands and fingers are detected and matched against a skeleton model (e.g., Microsoft Kinect). Segen and Kumar use a camera and a point light source to track a user's hands [171]. From the projections of the hand and its shadow, the system can obtain depth cues and thus calculate the position of the hand in 3D space.

Gesture recognition can also be combined with smartphones, and we shall explore this later in this chapter.

5.3.3 APPLICATIONS AND CASE STUDIES

While many researchers have created research prototypes, there are gesture-based public display systems that have been deployed in the wild. Notable examples of systems that have been used outside the laboratory are the StrikeAPose [193] and Looking Glass projects [137].

The StrikeAPose project created a game based on physics simulation to motivate passersby to interact. Passersby saw their mirror image on the screen and could then interact with the screen to manipulate virtual cubes. By placing the cubes in specific target areas, they collected points. This work was used to explore how to teach users new gestures. The Looking Glass project was described in Chapter 4.

5.4 MOBILE DEVICE INTERACTION

Smartphones typically provide users with stand-alone applications, communication capabilities and access to a wide range of cloud services. However, smartphones can also act as a gateway to devices in the local physical environment. For example, many new TVs can be controlled using custom smartphone applications. In this section we explore how smartphones can be used to interact with pervasive public display systems—moving beyond the living room to enable users to customize and control pervasive displays and digital signage they encounter during their everyday travels.

5.4.1 USING PHONES FOR PERSONALIZATION

In the future, content shown on pervasive public displays will be heavily personalized to reflect the interests of viewers. Smartphones provide an obvious mechanism through which viewers can express their preferences and communicate with nearby displays. Such mechanisms may provide the opportunity for longitudinal personalization over multiple encounters with a display (or displays) or may be more akin to short-term session descriptors, highlighting an individual and their interests for the duration of a single encounter.

Early work focused on the use of Bluetooth as a means of communicating with nearby displays. For example, in InstantPlaces [101] users could alter their device names to "tag" features of their identity and to link with their Flickr profile. The system built a longitudinal profile for users over time based on the presence history of their mobile device's Bluetooth MAC address. By comparison, the work reported as part of the e-Campus project [48] did not create a long-term profile, but simply allowed the use of Bluetooth device names as a control channel for requesting specific content to be shown on a nearby screen.

With the decline in the use of Bluetooth and the need to support more flexible forms of personalization, new systems are beginning to emerge. For example, the Tacita system [108] uses a smartphone application to collect user preferences that are then communicated to cloud-based

FIGURE 5.2: Using smartphones to personalize displays [108].

applications for presentation on nearby displays. The system identifies nearby using displays using a display map and calculates proximity to a display based on a combination of Wi-Fi-fingerprinting, GPS and Bluetooth device proximity (providing much greater flexibility than approaches based on a single proximity detection mechanism such as Bluetooth). Tacita also differs from similar systems for personalizing displays because it reverses the traditional pattern of announcements, requiring a user's mobile device to determine its proximity to a display (rather than the other way around) and because the personalization requests themselves are issued via a trusted application (i.e the application the user would like to see at the display). These two innovations help support personalization while preserving user privacy.

5.4.2   INTERACTION

In addition to being used as a form of identification, smartphones can also be used to add interactivity to modern public displays: allowing smartphones to be used as remote controls for large screens.

Inline with the abundance of technologies available on modern smartphones, these remote controls can take a wide range of forms. At their most basic, such systems can utilize the traditional communications functionality of the devices (e.g., SMS) to support text-based control languages or simply forward cursor/selection interactions with the mobile device's input mechanisms (e.g., keyboard, touchscreen) from the phone to the display [55]. Other systems take advantage of the wealth of additional technologies shipped with modern smartphones to allow complex interaction patterns: for example, work using a display tiled with NFC tags allowed smartphone users a range

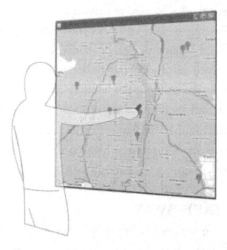

FIGURE 5.3: Reading display NFC tags through the smartphone [76].

of interaction operations including drag-and-drop and complex selection patterns (area selection, multi-selection) [76].

More novel forms of interaction include the touch projector that enables interaction with remote displays using a live video image on the mobile device [23] and using the flashlight of a phone and a display camera to create a novel pointing device [173].

5.4.3 CO-DISPLAYS AND CYBER-FORAGING

Used together, smartphones and public displays can act in a way that allows each to overcome the shortcomings of the other. A smartphone user struggling with the limited screen real estate of their device can "forage" for a nearby display and transfer their content for clearer personal viewing or to facilitate shared viewing. Such scenarios have been presented in a number of recent works [45, 150, 166, 202]. Pering et al.'s Elope system embedded operations (e.g., transfer a photo album from my phone to this display) in RFID tags—upon reading a tag, the smartphone receives the necessary information to form a Bluetooth network, expresses its intent and completes the operation. In cloudlet-based approaches [45, 166, 202], a user's smartphone acts as a transport mechanism for a virtual machine (VM). Upon encountering a display, the smartphone can trigger the transfer and instantiation of the VM to local computing infrastructure allowing the user to view and interact with their VM at a large public display. Commercially, the emergence of Miracast [198] and its integration with Android and other smartphone platforms provide a mechanism for a user to share content from their phone to a nearby TV, digital display or projector.

In contrast, smartphones can also be used as private co-displays to augment the view provided by public displays—thereby allowing viewers an "individual view" [113, 114], providing an uninterrupted view of display content and enhancing details that would otherwise be hidden or obstructed. The use of a smartphone as a private co-display in parallel with a public display also offers the ability to make aspects of the public display content deliberately private. On the public display, sensitive information can be blurred out, and then be made visible on a viewer's smartphone [15, 176].

In addition to using a smartphone's display, other output techniques are possible. For example, Rukzio et al. developed the Rotating Compass, which combines a public display and feedback via vibrations on the mobile phone to support pedestrian navigation [165].

5.4.4   INFORMATION TAKE-AWAY

Smartphones can also be used to support "take-away" functionality that allows viewers to collect information from the display—providing the digital signage equivalent of tear-off strips found on analog signs or click-throughs on conventional web advertisements. QR codes and RFID are two technologies that have ben widely used to support this type of functionality.

Researchers have continued to work in this area. For example, "Shoot & Copy" [22] used smartphone photos of portions of a display as a trigger for fetching content over Bluetooth or GPRS, while Digifieds [2] paired a display application with an Android application and used QR codes to allow the transfer of classified advertisements from the display to a smartphone.

More recently, She et al. [172] proposed using smartphone accelerometers to detect gestures made using the phone in order to indicate selection of items on a nearby public display.

5.5   ANALYSIS

There is an increasing trend toward more interactive forms of digital signage. However, there is no agreement on what form this interaction should take—for example should future displays be touch enabled, support gestures or enable interaction via mobile devices? Even within a broad technology area such as gesture interaction, there is no standard for the types of interaction and gestures that are possible. To avoid user confusion and to drive uptake of interactive displays, it is important that as our understanding of the best techniques develops we begin to converge on standard tools and techniques for user interaction in public displays.

CHAPTER 6

Systems Software

6.1 OVERVIEW

Modern display installations are highly complex systems consisting of a wide range of software components. Indeed, a common mistake made is to underestimate this complexity. For example, many small-scale users who are considering deploying a signage system often start by purchasing a screen and using a "retired" computer to drive the screen using standard office software such as a presentation package. Similarly, researchers often try and create their own signage system to support their research even when they have little direct interest in the software itself. For both users and researchers, however, the complexity of the problem space becomes apparent over time and such ad hoc systems are usually short-lived.

In this chapter we consider why developing the systems software necessary to support display systems is a non-trivial task and discuss some of the common approaches adopted.

6.2 UNDERSTANDING SIGNAGE SOFTWARE

The complexity of developing signage software can be attributed to a number of characteristics inherent in these systems. The list below isn't intended to be an exhaustive set of requirements for signage systems; rather, it highlights some of the characteristics that contribute to the complexity of signage software.

Complex functional requirements. Signage systems need to provide a wide range of functions including content creation, content distribution, content scheduling, real-time play out of content, control of specialized display hardware, analytics, remote management and monitoring and user interaction. This wide range of functional requirements inevitably results in complex systems. Even supporting one of these requirements (e.g., content distribution) is a significant undertaking in its own right.

Need to support a wide range of stakeholders and external systems. Signage software typically has to operate within a software ecosystem that may include a wide range of software tools such as content creation and management programs. The software also has to meet the needs of a wide range of stakeholders, including display owners, viewers, IT professionals and AV installers.

Reliable operation and embedded components. Failures in signage systems are, almost by definition, publicly visible. In many cases, failures can also have serious side effects (imagine, for example, the failure of a signage system in an airport). As a result, signage systems have high reliability requirements. This requirement is further complicated by the fact that many elements of the system may be embedded into physical displays and need to interface directly with a range of display and sensing hardware.

Remote management. Building on the theme of reliability and support for embedded components, many signage systems include displays that cannot be easily accessed for maintenance purposes. As a result, remote monitoring and management of displays is an important requirement.

Complex analytics. Signage operators are demanding increasingly sophisticated analytics that frequently include using a wide range of sensor technologies.

Situation dependent. Displays are increasingly expected to adapt to their environment and audience in order to provide more relevant content. Determining their situation and then deciding upon appropriate content to show is extremely challenging in many environments.

Cost sensitive. The signage industry is surprisingly sensitive to the cost of display hardware and software, with downward pressure on the cost of signage systems. This inevitably constrains the development effort that can be invested in individual software components—though this is somewhat mitigated by the increase in volume of sales.

In common with many similar complex systems (e.g., the GSM phone network), it is easiest to think of signage systems as a set of connected *segments*, each dealing with a specific set of concerns (Figure 6.1).

We are not aware of a standard segmentation for signage systems, so the divisions shown in Figure 6.1 are of our own construction rather than representing an accepted industry norm. Nevertheless, we feel that the segmentation we offer provides a good way of structuring our thinking about systems software for digital signage.

Content is central to signage systems, and the *content creation segment* is responsible for providing tools to enable the generation of appropriate content items. Once content has been created it needs to be ingested into the signage system, scheduled for presentation and distributed to the appropriate digital signs. This functionality is the role of the *scheduling and management segment*.

The most visible element of a signage system is the screen itself, and the *display segment* is responsible for all of the local operations necessary to present the required content on the screen.

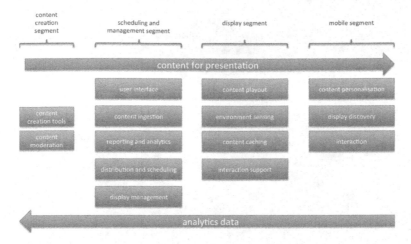

FIGURE 6.1: Understanding the architecture of signage systems.

In addition, in modern systems the display segment may also include a wide range of sensing and interaction functionality as it typically encapsulates all of the software that must run local to the display.

Finally, the *mobile segment* comprises any software that viewers run on their mobile devices to support interaction with a pervasive display. In many signage systems this segment is empty or consists of simple off-the-shelf software such as a QR-code reader. However, this is likely to change as signage software evolves to a richer model of pervasive displays.

While traditional digital signage systems can be segmented as shown in Figure 6.1, we believe that the trend toward more open systems is likely to lead to an increasing amount of content being sourced from viewers close to displays. As a result, we believe that Figure 6.2 provides a more likely depiction of the relationship between segments in future display systems.

This represents a significant shift from a world in which content flows downstream from content producers to viewers (and analytics upstream from displays to content producers) to a more complex eco-system in which content, scheduling requests and analytics may be sourced from multiple points, both upstream and downstream of displays themselves.

6.3   SOFTWARE AS A SERVICE AND THE ROLE OF THE CLOUD

Traditional signage systems consisted of dedicated software installed on computers connected to displays (or indeed embedded in them) plus associated software for content creation and management. With the emergence of the Web and, in particular, with the ability to show high-quality video

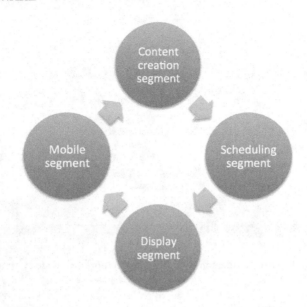

FIGURE 6.2: Understanding the architecture of signage systems.

within a browser, many signage systems moved to a thin-client model. In such systems the display nodes are relegated to running a web browser that connects to the signage system.

There are a number of design trade-offs involved in moving to a thin-client approach. In general, these trade-offs mirror the standard arguments for and against thin-client architectures. For example, a significant advantage is that using thin-clients makes it much easier to support a wide range of display end-systems since clients need only support a web browser to access the full range of signage functionality. Moreover, since much of the content that signage systems access these days is actually live content obtained via HTTP, support for web technologies is required in any case.

However, thin-client systems also come with disadvantages—for example they are less able to survive periods of network disconnection (even when utilizing the capabilities of modern browsers). One key difference between signage systems and most thin-client applications is of course that while the output of a signage system is highly visible there is often not a user (in the traditional sense) actually interacting with the display. As a result, any errors need to be detected and remedied automatically or by a remote system administrator. This further complicates the process of assessing the benefits of a thin-client approach that, in the end, may be driven as much by architectural preference as hard requirements.

Recently, commercial signage systems have begun to deploy signage systems using a model of *software as a service (SaaS)*—e.g., the BroadSign Digital Signage system. Such systems remove the need for signage operators to install dedicated software for managing their content and displays;

instead, this is all carried out by software in the cloud. A SaaS approach would classically be combined with a thin-client–based display but this is not essential. Indeed, it is perfectly reasonable to envisage a display system in which management and content processing are carried out in the cloud using a SaaS approach while displays themselves are driven by dedicated hardware/software solutions that leverage the cloud but also provide significant local capabilities though dedicated, installed software.

In the following sections we consider three of our four architectural segments in more detail (since mobile devices are primarily used for interaction a full description of the software that can be deployed on mobile devices as part of the mobile segment was presented in Section 5.4).

6.4   CONTENT CREATION SEGMENT

Our first segment is concerned with the creation of content for display. In many cases, the tools used for this purpose are not signage-specific but are generic tools used for creating content such as images, animations, videos or soundtracks. Such tools are appropriate for professional content creators and we do not discuss these further.

However, most signage systems also recognize that in many cases content will need to be produced quickly and cost-effectively by non-professionals and hence provide simple tools for creating straightforward content items. For example, Sony's Ziris system [180] includes the Ziris Create software tool that enables users to create content for their signage network. Researchers have also attempted to create simple interfaces for content creation. These interfaces range from "wizards" that guide users through the creation of posters or announcements to systems that provide a replacement printer driver and enable users to publish to a digital sign simply by printing a document from any of the standard applications with which they are familiar.

In the research domain there has also been significant experimentation with supporting user-generated content. An example of such a system is Dynamo [94], which supported cooperative sharing and exchange of media—allowing viewers to interact with the display and to leave media messages for future viewers. User-generated content has also been explored as part of the "display personalization" agenda by researchers such as Jose [101] whose system allowed users to send messages to displays from their mobile phones. The messages were subsequently shown on the displays allowing users to personalize their environment and helping to build audience engagement with the displays.

User-generated content can also take the form of specific content items such as advertisements. In the Digifieds system, for example, users are allowed to post classified advertisements onto public displays (see Chapter 7) .

Even when users are allowed to generate content there is often a shortage of high-quality, relevant content to show on a public display. This is a significant problem because a lack of content leads to a perception that the display is "out of date" or "stale" and encourages users to ignore the display. To try and address this problem, Memarovic and his colleagues proposed the idea of autopoiesic content [126], i.e., self-generative content that is automatically created. In the system they have prototyped, the authors endeavor to create interesting and relevant content in the form of "fun facts" by matching local context information with regular scheduled information to produce information snippets that are highly localized. The authors cite the example of creating a new fun fact by taking the number of people around a display and matching it with a population figure in its content fragment database: "The population of Pitcairn Island (50) is five times more than today's average number of people at the display (10)" [126]. While still at an early stage, this avenue of research is especially exciting because it offers the potential to ensure displays are constantly supplied with fresh content—helping to increase their perceived value.

6.5   SCHEDULING AND MANAGEMENT SEGMENT

6.5.1   CONTENT SCHEDULING

The scheduling and management segment provides the back-end for a signage system—it is essentially responsible for mapping content items to presentation slots on displays. To achieve this, many systems provide support for two abstractions: *display groups* and *content playlists*. Display groups are simply sets of displays that can be treated as a single unit for scheduling and management purposes. In practice, this means that these displays will show the same content. Playlists typically correspond to collections of content along with a specification of when the content should be shown. The scheduling and management segment allows signage operators to create display groups and playlists and then to schedule playlists onto display groups.

To support the above functionality most signage systems provide a user interface based on some form of timeline similar to that found in many video-editing packages. Users can drag content items onto the timeline to create a playlist that can then be scheduled.

Historically, this approach was adequate for most signage installations. However, the emergence of interactive signage systems has significantly complicated this segment. For example, instead of playlists that are focused on simple linear playout of content, playlists must now support content scheduling that is conditional on external factors such as audience presence or interaction. Futuristic scenarios such as display appropriation further complicate scheduling.

Our experiences suggest that even without this additional complexity users find scheduling content to be a demanding task [43]. As a result, in one of our own systems (e-Campus) we have focused on providing users with a very simple user interface based on the idea of "channels." The key design insight for this interface was to separate the roles of content providers and display owners:

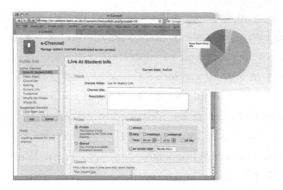

FIGURE 6.3: The e-Campus content scheduling user interface.

1. Content providers generate content (images, videos, web pages, live video streams) and organize it in logical containers called "channels." A content provider has full ownership of their channels and can add and remove content and set the date and times it is available. Content providers do not necessarily control displays and they have no say on where or when their content is displayed. Instead, channels can be shared and their content scheduled to a display after its owner subscribes the display to the channel.

2. Display owners control one or more physical displays, typically in a "local" physical space for which they have oversight. A display owner controls their displays' channel subscriptions. A display can be subscribed to zero or more channels. The display owner can also set the time at which their displays turn on and off.

Content providers use a web interface (Figure 6.3) to create and manage channels. Each channel also gets an associated network file share so that content can be managed conveniently via drag and drop. A web interface is also provided for display owners to control their subscriptions. Users who are both content providers and display owners can create "private channels" only available for their displays.

The system produces a schedule for each display based on the combined set of content items from all its subscriptions. We weight all channels equally, although newer or less played content, and specific types of content (such as emergency alerts and interactive applications) are given higher priority.

While this type of approach provides users with a very simple interface, it requires the system itself to be able to make sensible scheduling decisions and this is not easy. In [186] the authors discuss the key challenges in content scheduling in future pervasive display networks. They identify a wide range of scheduling requirements, including showing content at specific times, showing content a specific number times (for example to meet commitments made to advertisers), ensuring content items are shown consecutively (or, conversely, that content items are always separated by a minimum time interval), showing content as a result of contextual triggers or sensor events and supporting content rescheduling based on user interaction.

The decision on when to schedule a specific piece of content can depend on a wide range of factors. In [186] four distinct factors are identified: display owner preferences, display viewer preferences and presence, content availability and the display's context. Other example factors might include content developer preferences or financial or contractual constraints. For a detailed discussion on scheduling issues, see [186].

Once content has been scheduled the content items need to be distributed to the displays. In thin-client systems this may be trivial, but in systems that support local caching more sophisticated content distribution mechanisms may be employed to ensure that the content is available on the display node when required. For example, many display networks deployed in retail situations download new content overnight as a batch job in order to ensure that the content is available locally when required.

6.5.2   CONTENT INGESTION

So far in our discussion of the scheduling and management segment, we have assumed that the content has already been ingested into the system or is trivially available. While ingestion has traditionally been a fairly simple task in signage networks, the shift toward more open display networks is making this more challenging and new ways of ingesting and distributing content are being researched. One example of how this might be achieved in the future is through the introduction of a pervasive display equivalent of an "app store" [44]. Application stores offer the possibility of opening up display networks to content and applications from a range of sources. While an application store for public displays has the potential to do what such stores did for mobile devices—to open up a "closed" model of development and deployment—there are significant differences between the mobile device market and that of public display networks. Foremost is the set of stakeholders; mobile devices are typically owned and managed by a single individual, with third parties providing communication services. In contrast, display networks feature a stakeholder set including display and space owners, viewers (users) and content (application) providers. In many cases, the owner of the display is different from the viewer (e.g., an advertisement outside a shop is managed by the shop owner but the content is for potential customers).

The existence of these many interests adds a layer of complexity to applications as they emerge to benefit different stakeholders. We discriminate between two classes of applications:

Applications that primarily benefit the display owner. For some display owners (e.g., a small business owner), creating high-quality content is a significant problem. Utilizing a public display application store to purchase applications appropriate for their displays would reduce the burden of acquiring content. Sample applications might include generic or localized advertising, coupon generation, information delivery and even interactive jukeboxes to contribute to the environment.

Applications that primarily benefit the display users. At present, viewers of public displays often have no control over the content. We envisage a world in which members of the public purchase applications reflecting their own interests for display on one or more public displays. Sample applications include sports results, interactive city guides and applications that allow users to make an artistic statement by showing a particular picture when they are near to a display.

While applications in this second class principally offer benefit to display users, a display owner may choose to provide the run-time environment of a display because, for example, they feel the resulting attention given to the display may carry over to other content. In both cases, application developers may benefit though financial exchange, reputation or data gathering.

The business models underpinning an application store for public displays are also likely to differ from those of similar stores for personal devices. Current business models for display networks are often advertising focused, with companies maintaining large networks and selling the space to those wishing to advertise (e.g., [170]). The provision of an application store for public displays, and the consideration of stakeholder cost/benefit exchanges, raises design considerations regarding the nature of an appropriate business model. Such a model will need to consider how to reconcile the conflicting demands of stakeholders, who should pay whom for the applications (and with what?), and how such payment models impact on the acceptance of applications that primarily benefit the user by display and space owners.

The introduction of components such as app stores also further complicates scheduling in pervasive display systems. In mobile phone–based application environments the decision to start an application is made by the user of the device. In conventional display networks the decision to schedule content is made by the display owner. In a display network with application stores the situation is significantly more complex and there are other possibilities. The display owner may buy applications from the store, in which case they will still expect control over application scheduling. Alternatively, applications may be purchased by users who expect them to show as they pass by a display. In addition, application creators may wish to impose constraints on when particular content can be displayed. Providing appropriate interfaces for scheduling control is an important issue that must be addressed, and the appearance of "control" over the display is desirable for almost all stakeholders in the display scenario.

6.5.3   DISPLAY MANAGEMENT

We note that in addition to content scheduling and content distribution the scheduling and management segment is normally responsible for display management and reporting. Display management typically includes functions such as remote status monitoring, controlling power states and software updates. These functions are critically important because displays are often deployed in areas that are difficult to physically access or where there are no local support experts. As a result, many administration tasks need to be conducted remotely. Indeed, it is perhaps most appropriate to think of most public displays in the same vein as embedded systems.

It is worth highlighting that status monitoring is a particular challenge for public displays. This is because the only thing that is *really* important is what is on the screen of a public display, yet this is the most difficult thing to monitor. Monitoring software that reports on the status of the player hardware is typically of limited value because this only provides an indication of the state of the computer driving the display, not the display itself. Indeed, even when the display itself is instrumented to measure, e.g., power state, this is still only of limited value—if, for example, the display surface itself is damaged, then the system may not be displaying content even though no errors are being reported.

Reporting will often include basic audit trails and analytics of content impressions that can be used to support billing and contractual requirements. Collecting such reports is problematic because of the issues described above so it is necessary to make "reasonable assumptions" when examining this type of data. Indeed, the area of signage analytics is especially challenging because what is easy to capture, i.e., the number of times a piece of content is shown, is much less relevant than information on the number of people who viewed a specific content item. The state-of-the-art is to use cameras and image processing techniques to count the number of viewers who look at a display.

6.6   DISPLAY SEGMENT

6.6.1   KEY FEATURES

Content and schedules produced by the scheduling and presentation segment need to be presented on physical displays, and this is the role of the display segment. In addition, we have allocated support for sensing and interaction to the display segment as this functionality is often implemented locally to the display.

In the case of a thin-client the scheduling and presentation of content may well be trivial as the display node will simply run a web browser with the content being pushed from the signage system back-end. In systems where display nodes have more local autonomy some scheduling decisions may be carried out locally, for example responding to local sensor events, and content may be sourced from a local cache. In such systems a wider range of presentation options may also be possible. In particular, such signage players may use dedicated software components or renderers to present output to the display. This would be appropriate if, for example, a display needed to show the output from a program that could not be easily rendered within a web browser.

Whichever approach is used, display segments also require the standard features expected in an unattended system such as watchdog timers to ensure the system is restarted in the event of problems with content items.

The display segment may also be responsible for coordinating display across multiple displays if, for example, the installation takes the form of a video wall. Synchronization of multiple displays for such installations has been extensively researched within the graphics community, leading to systems such as SAGE (Scalable Adaptive Graphics Environment) [98], and tight synchronization between displays is a topic in its own right and beyond the scope of this lecture.

To enable them to react to their environment, modern signage systems can be equipped with a wide range of sensors. These can range from simple movement sensors that automatically switch displays off in unoccupied spaces or trigger attract sequences through to sophisticated image processing systems that can tailor content to the age and gender of the viewer. In general, interfacing to these sensors is carried out by local software, though the output from this local processing may be streamed to the back-end to drive content changes.

Displays also need to support interaction. Once again, a wide range of options are possible in this space. For example, in the case of simple touch screens input may well be mapped locally to mouse events that are then passed to the relevant signage application. However, more sophisticated forms of interaction such as gesture recognition may require alternative mechanisms for routing the output of the interaction component to the appropriate signage application. Several researchers have proposed toolkits for supporting generic interaction with signage systems (e.g., PuReWidgets [33]). These toolkits typically allow users to interact with a screen using a range of technologies such as touch, gestures or SMS commands and then map these interactions into a single input stream that is delivered to the signage application.

6.6.2   AN EXAMPLE SOFTWARE PLAYER

As an example of a modern signage player, we now describe the main features of Yarely, a signage player developed by the authors to support open pervasive display networks.

Yarely obtain instructions on which content to play in terms of *Content Descriptor Sets* from components in the scheduling and management segment. The purpose of the Content Descriptor Set is to provide a description of a set of content items to be played by the node, the circumstances in which they should be played and the location of any required media. We term any system component that creates Content Descriptor Sets a *Descriptor Factory*. Communication between the Display Node and Descriptor Factory (DF) occurs through the transmission of a Content Descriptor Set from the DF to the Display Node. A Content Descriptor Set is encoded using XML (Figure 6.4), and, from a design perspective, transmission of the Content Descriptor Set is protocol-agnostic. Each Display Node may use Content Descriptor Sets from multiple DFs and each DF may serve many nodes.

```
<?xml version="1.0"?>
<content-set name="Visit Day 2012" type="inline">
  <auth handler="none"/>
  <feedback/>

  <constraints>
    <scheduling-constraints>
      <playback order="random" avoid-context-switch="false"/>
      <time><between start="09:00:00" end="17:30:00"/></time>
      <date><between start="2012-09-22" end="2012-09-22"/></date>
      <priority level="medium"/>
    </scheduling-constraints>
  </constraints>

  <content-item content-type="video/mp4; charset=binary" size="42026281 bytes">
    <requires-file>
      <hashes><hash type="md5">f421a52be576891e0948a07067b8dc38</hash></hashes>
      <sources><uri>http://content.com/vd2012-40Mb.mp4</uri></sources>
    </requires-file>
    <constraints>
      <scheduling-constraints>
        <preferred-duration>246.29</preferred-duration>
      </scheduling-constraints>
    </constraints>
  </content-item>

  <content-set name="7: Visit Day 2012" type="remote">
    <requires-file>
      <sources>
        <uri refresh="2 minutes">http://content.com/channel.php?id=202</uri>
      </sources>
    </requires-file>
  </content-set>
</content-set>
```

FIGURE 6.4: A sample Content Descriptor Set.

A Content Descriptor Set is comprised of a combination of two main types of element: content-sets and content-items. A `content-set` is essentially a container element for content-items and other content-sets, while a `content-item` describes a single item of media that may be rendered at a display.

Yarely features a component-oriented design and is divided into five key modules (Figure 6.5):

Subscription management. The subscription management module is responsible for maintaining connections with the Descriptor Factory for the purpose of requesting and receiving the Content Descriptor Sets that describe the subscriptions. It polls periodically for changes in its subscriptions and passes any changes on to the playlist generator and scheduler.

Sensor management. The sensor management module reads data from sensors and environmental sources (e.g., context, user presence and interaction) that may trigger injection of interactive content onto the node.

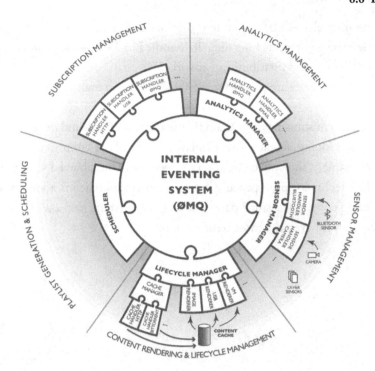

FIGURE 6.5: Node architecture.

Playlist generation and scheduling.   This module processes the available subscriptions to derive a playlist of content items to be played in the immediate future and is responsible for the scheduling of these content items onto the screen in response to changing circumstances (e.g., time, sensing events) and in line with the constraints specified in the Content Descriptor Set.

Content rendering and lifecycle management.   Content rendering refers to the physical representation of content items onto output hardware (e.g., display, speakers). The lifecycle of such content items includes the caching of media at the display node, the preparation of content in advance of playback, playback and cleanup on completed playback.

Analytics.   The analytics module is responsible for returning data to external servers (e.g., the Descriptor Factory or application provider). These may take the form of summary statistics or notifications on a per-event basis.

Each module is composed of a managing element plus a set of further plug-ins that handle specific software or hardware requirements. The module for Content Rendering and Lifecycle Management, for example, relies on a set of renderer plug-ins needed to handle content types, including web pages, images and pre-recorded and live-streamed video. The renderers are platform specific and are the only element of the system that needs to be adjusted from system to system. Within the lifecycle management, caching plug-ins handle HTTP download and other mechanisms, e.g., bit torrent or other content distribution networks.

Yarely is simply one software player for digital signage and many others exist offering a wide range of features.

It is, however, worth stressing once again that many researchers and developers underestimate the complexity of developing signage and display systems. This commonly occurs because the principal function—namely, showing content on a screen—can be achieved straightforwardly using existing off-the-shelf technologies, especially if a web-centric approach is adopted. However, as we have shown, rendering content is a small part of any signage system, and a wide range of additional capabilities are required for even the most basic system.

CHAPTER 7

Research Tools and Techniques

7.1 INTRODUCTION

Pervasive displays represent a young and exciting area of research. In many areas of computer science there are well-accepted research tools and techniques that are used to help answer common research questions. For example, most networking researchers will be very familiar with packet sniffers and network simulation tools such as ns-2/ns-3, while database and file systems researchers will have used a wide range of benchmarking tools on standard data sets. For researchers in pervasive displays, however, the choice of tools and techniques is much less obvious—there are no widely accepted test data sets, tools or techniques.

Indeed, researching pervasive display systems is extremely challenging for a number of reasons. First, there is often *no single goal* that pervasive displays (or their content) try to achieve. Ads most likely strive to maximize public attention, interactive games may want to create an engaging experience, informative applications such as a public transport schedule may aim at maximizing usability and some displays may be designed to fade into the background, just presenting ambient information. Hence, metrics for display systems need to cope with different content, situations and purposes if meaningful comparisons are to be made.

Second, pervasive display systems often need to be evaluated in the wild because there are *no models or simulations* that can be used for experimentation. The closest signage comes to models are static models of attention such as the Fraunhofer frequency atlas for Germany that provides an attention estimate for streets, public transport infrastructure and pedestrian areas [121]. The model considers different factors, including points of interest (restaurants, recreational facilities), population density and structure, traffic, street categories and socio-demographics as well as socio-economic data. While this model is useful for estimating likely viewer numbers for a given screen, it says nothing about how such views might respond to a given display or, indeed, the impact the display itself may have on the environment.

Finally, we note that *conducting field studies or deployments of pervasive display systems is extremely challenging* in terms of the investment required to make these a success. In particular, issues ranging from ethics requirements through engagement with multiple stakeholders to developing prototypes that can operate unattended for extended periods all add significantly to the time and financial costs of conducting research in this space. However, the potential impact of public display research on the everyday lives of millions of viewers makes this topic well worth the investment.

In this chapter we aim to assist researchers starting out in this area by providing an overview of common research approaches used in the field of pervasive displays.

7.2 EXAMPLE RESEARCH AREAS

Given the relative immaturity of the field it is perhaps unsurprising that researchers are often trying to answer a very wide range of questions when they experiment with pervasive display systems. However, it is notable that there is very little systems-oriented research in this field; instead, researchers are more often focused on trying to understand the impact of displays on users in areas such as user experience, user acceptance, privacy concerns and social impact. In the following list we provide more details on a number of example pervasive display research areas:

Audience behavior. A major research focus is often on how an audience behaves around a display. This topic is particularly interesting for application developers and content creators as novel interaction technologies and techniques emerge. For example, showing content on a planar screen may lead to very different behavior when compared to showing the same content on a cylindrical screen due to the lack of framed content [17]. Researchers have studied a number of systems in this area and observed effects such as the Honeypot [31, 137] (interacting users attract more users), the Sweet Spot [16] (a preferred position in front of the screen) or the Landing Zone [137] (people only realize that a display is interactive after passing by). Though mostly conducted in the real world, there are also examples of lab studies [16]. Audience behavior is most commonly evaluated through observations [60] and log data [163]. For a fuller discussion on audience behavior, see Chapter 4.

User experience. User experience depends on different features, including content, presentation, functionality, and interaction [19]. A good user experience may lead to a higher motivation to use the application and possibly draw the user in for as long as possible. For example, if a display application makes the user look "good," this staging effect is likely to increase interaction times. Researchers have examined different interaction techniques and their effect on user experience based on (standardized) questionnaires, e.g., interactions mediated through a mobile device [10] or direct touch [160]. Interaction is closely related to user experience—see Chapter 5 for more details on this topic.

User acceptance. User acceptance investigates users' motives and incentives for using a display. For example, passersby may not use or even look at a display if there is no perceived benefit, as is the case with many advertising-based displays today. User acceptance can be assessed qualitatively based on subjective feedback, e.g., in focus groups to collect the target group's view and concerns [40] or quantitatively based on questionnaires [128] or interaction logs.

Privacy and security.   As displays become equipped with an ever increasing number of sensors they are prone to raising privacy concerns of the user. The video feed from a display could, for example, be used to identify the user in front of the display and hence create movement profiles across display networks. A similar problem arises as displays enable user-generated content that requires personal information such as an email address to be provided. A number of projects have explored this topic, including Alt et al., who looked at how mobile phones can be used to help overcome privacy issues [2] when exchanging personal information with a display, and Shoemaker and Inkpen, who explored a novel interaction technique that allows private information to be shown on a shared display [176].

Social impact.   Pervasive displays can have quite a strong social impact. Examples are displays connecting communities around the display [2, 40] or displays showing content that serves as a catalyst for social interaction, e.g., by engaging users to play together [137], take pictures together, or simply talk about the content [126]. Representative projects include the CityWall, a large display in Helsinki that allows multiple users to browse photo content on flick and investigates roles of the users and social configurations [148], and FunSquare, an application that uses contextual information about a place to create fun facts, thus trying to make passersby discuss the content [126].

User performance.   Many displays aim at minimizing interaction times. For example, a display providing travel information at a train station should provide an interface that allows the user to find the required information as easily and as quickly as possible. Examples of studies of user performance include comparison of direct touch at the display and the mobile phone [5] and in-depth investigations of interaction techniques based on cameras, mobile phones [10] or direct touch [47]. User performance is typically quantified by measuring task completion times and error rates [10, 47, 168]. Despite not being portable, displays offer a number of benefits in terms of scale and immediacy that can lead to high levels of user performance in many cases.

Display effectiveness.   Defining and measuring the effectiveness of a display is challenging. However, in a commercial context owners of displays often want to know how many views a display attracts and how much revenue a display generates and this can be considered some form of measure of effectiveness. While commercial audience measurement is still in its infancy, there are examples of research into how to measure the number of people passing by a display [138], looking at it [86, 138] and finally starting to interact [137]. The issue of sign analytics was discussed in Chapter 6.

7.3   FUNDAMENTAL RESEARCH CHOICES

Answering research questions such as those described above requires many different study types and methods. At the outset, researchers need to identify whether their research is descriptive, relational or experimental.

Descriptive research aims to accurately describe what is going on in a certain situation using techniques such as observations [138], interviews [2], focus groups [4] or photo logs [2]. In descriptive research multiple prototypes for comparison are not needed. It is striking that the vast majority of public display research includes descriptive methods. Good examples are the CityWall [148] and Worlds of Information [95]. In both studies, a single prototype is deployed and user behavior around the display is measured, analyzed and described. Descriptive research is especially suited for a research field like public displays that is in an early phase of development and does not yet have general theories that need to be evaluated. However, descriptive studies of single prototypes create isolated points in the pervasive displays design space, and it is difficult to compare results and designs or to generalize results. This makes it challenging to ultimately understand the structure of the entire design space, and hence the progress of public display research may be hindered, in the long run, if it continues to focus purely on descriptive studies.

Relational research aims at showing that two or more variables covariate, i.e., that there is a relation between two or more factors. In relational research causality is often not attributed, i.e., it is unknown which of the variables causes the other to change or whether both depend on a third, unknown variable. Relational studies are rare in public display research, in part because not many relationships between different dependent variables are known or considered to be interesting. Exceptions include ReflectiveSigns [138], where it is shown that the time people spend looking at public display content does not correlate with people's stated interest.

Experimental research aims at determining causality, i.e., that one variable directly influences another variable. Experiments typically possess the following characteristics: they are based on hypotheses, there are at least two conditions, the dependent variables are measured quantitatively and analyzed through statistical significance tests, they are designed to remove biases and they are replicable [112]. Experiments can be conducted in the lab (more control) or in the field (higher ecological validity). Whereas a lot of experiments have been conducted in the lab (e.g., in order to evaluate user performance with regard to a novel interaction technique [10, 23, 47, 104]), real-world experiments are rare in public display research because of the time and costs involved. Experiments in which multiple variations of a prototype are developed for comparison purposes are particularly uncommon, with a notable exception being the Looking Glass project [137] that tested the influence of different interactivity cues on the number of people interacting with displays.

7.4 COMMON APPROACHES

7.4.1 DOMAIN ANALYSIS

A range of techniques can be employed early in a research project's lifecycle to help develop an understanding of the problem domain.

Focus groups vary in format but are usually run with 5–8 people in sessions lasting about 1–2 hours, including a demonstration of the system (e.g., a novel interaction technique for displays) and hands-on trials, followed by a discussion. The discussion is typically led by one of the researchers. A good example is the work of Cheverst et al., who use focus groups to discuss multiple system designs with different degrees of interaction [40]. The advantage of focus groups is that feedback can be obtained at very early stages of the design process.

Interviews in pervasive display research are often semi-structured, i.e., the interviewer follows some pre-defined guidelines but is able to explore topics in depth if they obtain interesting responses. Interviews can also be conducted in context (e.g., shortly after the subject used the system) and are a powerful method for understanding the user's views (e.g., concerns, problems and opinions [2]). *Questionnaires* have been used in many of the reviewed projects in order to assess, e.g., user experience [16], user performance and the users' views [10].

In *ethnographic studies* settings are investigated without intervention, e.g., without deploying a prototype. Ethnographic studies have been used extensively to inform the design of public display systems. Alt et al. conducted an ethnographic study to assess the motivation and intentions of stakeholders as well as social impact [2]. Huang et al. investigated current practices around public displays [86].

7.4.2 LAB AND FIELD STUDIES

Lab studies aim at evaluating a system within a controlled environment. Lab studies can be descriptive, relational or experimental (see Section 7.3). During the lab study both qualitative data (e.g., interviews [137] and observations [16]) and quantitative data (e.g., task completion time and error rates [10, 23]) can be collected. The advantage of lab studies is that external influences (such as other passersby, environmental conditions) can be minimized and (sensitive) equipment for proper measurements (such as cameras and sensors) that would be difficult to deploy in public can be used [16]. The disadvantage of lab studies is that they may provide only low ecological validity and that the dynamics of the real world are excluded.

In contrast to lab studies, field studies aim at evaluating a system or technique in a (semi-) public setting. In contrast to deployment-based research, they are rather short (days to months) and focused on a single research question. Similar to lab studies, they may be descriptive, relational or experimental. Data collection in the field is often cumbersome and time-consuming, since automation may be difficult due to privacy issues (e.g., recording video in a public space). The advantage, however, is that a high ecologic validity of the data can be assumed. Furthermore, there are aspects such as effectiveness [138], social effects [126], audience behavior [137] and privacy implications [2] that are almost impossible to measure in the lab. The disadvantage of field studies is that they are usually complex due to the high number of potential influences and require a tremendous effort in preparation (finding a suitable place, legal issues, etc.).

7.4.3   DEPLOYMENTS

Deployment-based research introduces technology (e.g., public displays) into a specific setting (e.g., a city). User feedback and involvement are obtained and, in an iterative process, the deployment is improved. At the same time, this data is used to build and refine the theories, which in turn generates new research questions that can be addressed through changes in the deployment. In contrast to field studies, deployments are integrated into the everyday life of their users. While having a high degree of realism, a limitation of deployment-based research is that interpreting results becomes challenging because of the many confounding factors that impact on the results obtained.

Deployments form part of a continuum from lag studies through cultural probes and technology probes to deployments [88]. *Cultural probes* support users with things like cameras to document their lives, while *technology probes* introduce small prototypes in order to understand a given domain, sometimes without the scientific rigor introduced in experiments. Only *deployments*, however, really become permanent useful artifacts in everyday life. Deployments enable researchers to investigate longitudinal effects of use that cannot be investigated with other means. They are also the only method that can eliminate the novelty factor that influences the results obtained via other kinds of studies. However, the maintenance of such deployments binds considerable resources. Examples of deployment-based research are Hermes [38], the Wray display [40], e-Campus [183] and UBIOulu [145].

Quantitative data is often captured as part of deployments (and to a lesser extent field studies) and is particularly helpful when collected over a long period of time. Tools for data logging include sensors that track motion [16], eye gaze [169], presence [137] and user interaction [2]. Quantitive data is used in many studies to help describe, e.g., the trajectories [16], time of day [2] and type of content [169] with which interaction occurred.

Observations also provide a powerful tool to analyze audience behavior around displays [16], their effectiveness [137] and social impact [95]. In general, two forms of observations can be distinguished: automated and manual. During *automated observations* users are observed by cameras installed in fixed locations (potentially filming both the screen and the viewer) [16]. The video footage can be analyzed post hoc using computer vision methods such as shape or movement detection, eye recognition or manual annotation and coding [16]. When conducting *manual observations*, data is gathered by observers, e.g., by taking field notes or pictures and videos from both the subject and the display [137]. In this case, the observers usually hide in a location from which both the screen and the interacting persons can be seen [2]. The advantage of this method is that users behave most naturally if they are not aware that they are under surveillance, making the findings highly ecologically valid. Of course video-based observation may compromise privacy, and it may be very difficult to draw conclusions as to why users behaved in a certain way. Therefore, observations are often combined with (post hoc) interviews.

7.5   GUIDELINES FOR RESEARCHERS

7.5.1   GENERAL POINTS

There are few hard and fast rules for research in the field of pervasive displays. However, we believe that the following points should help those considering undertaking research in the field.

Know whether the study is descriptive, relational or experimental. It should be very clear whether the study is an experimental one or not. If there is no theory or good hypothesis, as is often the case with public displays, a descriptive study is likely to be more suitable. Experiments, however, help establish causal relationships between phenomena and allow the design space of public displays to be structured in the long term so are particularly valuable.

Choose your focus on internal, external or ecological validity. Oftentimes, control, generalizability and realism cannot all be achieved at the same time. It is important to make clear which kind of validity will be the focus and which validity will be partially sacrificed. In public display research, ecological validity is typically more valued, but it has to be clear how internal and external validity are impacted, and what measures can be taken to improve these (e.g., randomization to decrease the influence of confounding variables).

Consider the impact of the content. Public display research is not possible without content, but the impact of content and other factors on usage is indistinguishable [183]. Thus, every study is at risk of producing results that are only valid for the particular kind of content tested. Testing different contents might help.

Understand the users and the environment. Public displays may have vastly different users at different locations and different times. In [137], for example, school children in the morning behaved very differently from people at night, while at one location very few people looked at the display because they turned their head in the opposite direction to look down a road as they approached it. Deployment in public space also introduces ethical issues, and anonymization of any required data will usually be necessary.

Be prepared for small sample sizes. One of the most common problems with public displays is that they do not receive a lot of attention. This is particularly problematic for interaction studies because passersby do not expect displays to be interactive [137] nor do they expect to find interesting content on the displays [138]. Engaging people who actively promote displays can help to raise audience awareness of a display's interactive capabilities [144], and creating highly customized content that reflects the users' interests may also help but is expensive [183].

7.5.2   CONSIDERATIONS FOR FIELD STUDIES AND DEPLOYMENTS

In addition to the general points discussed above, field studies and deployments need particular care. First and foremost the system and software created must exhibit a degree of robustness that far exceeds that required for traditional lab-based work. In addition, the points below are relevant and are based on many years experience of conducting this type of research (see [67] for more details).

Engaging with stakeholders. Deploying systems for people to use is always a costly process. Designing a system that meets expectations, helping set those expectations and ensuring that no health, safety or regulatory constraints are contravened takes great care. We recommend actively seeking to involve the various display stakeholders and learning as much as possible from them about their concerns and constraints. Care should be taken in setting expectations about the system, especially early on: in situ testing can be confused with a live system, and periods when the system is deliberately off, with system failure. Such testing cannot be avoided: it is invaluable for understanding the practical, environmental and operational concerns of running your system in the wild. Researchers carrying out field studies and deployments also need to prepare themselves, their team and their work for public scrutiny—there will be curiosity about the system, and you may be held accountable to funders, media or the public.

The importance of good content. As we found, and as can be seen more generally from many other public-facing ubicomp systems, such as GUIDE [36] and Uncle Roy All Around You [14], where content is intrinsic to the system, good content is essential. For many users the content *is* the system, and poor content results in a poor user experience and reflects badly overall. Considering content too late in the project life cycle can lead to a mismatch between the needs of content and its creators, and the facilities offered by the system. Generating compelling content is a non-trivial task requiring specialist training and toolchain knowledge. In our experience it is normally best to go outside the project team, often to art and media professionals, and resource this appropriately.

After deployment comes maintenance. Issues affecting maintainability will differ for each system. However, we note that for public deployments it is crucial to ensure adequate physical access to, and control of, the space in which the system is deployed. To pick a somewhat extreme example: for one of our own installations a series of projectors, which we had perhaps hastily assumed would require little attention, were located above a road in a bus station. Repairs necessitated closing the road, rerouting the buses and assembling a sizeable scaffolding tower—each repair took on average five weeks to achieve as a result! For long-lived systems we also suggest that—rather than, as is the temptation, creating a deployment in the lab and then moving this into the field—a duplicate set-up should be kept in the lab to ease testing of proposed maintenance procedures. We believe that in most cases the extra costs involved will be more than justified.

Remote observation and control. Researchers should ensure that the final output of the system can be remotely monitored and controlled where possible. Pervasive display systems are often complex conjugations of hardware and software, and even with taps into remote communication channels and command shell or virtual desktop access to parts of the system, it is seldom easy to appreciate what the user is actually experiencing, let alone make good guesses as to what is and isn't working correctly. We have found that building "white box" components that externalize their internal state, or send periodic status announcements, e.g., via a publish-subscribe event-channel or tuple spaces [99], can make observing, debugging and even extending the system much easier. Anticipating and planning for (and not underestimating!) the need to restart systems and components, fault recovery and managing the flow of content and growth of log files, simply or automatically, will pay off in the long run. Adjunct to this, remote networked cameras and microphones, while seldom popular with end-users, do make fault reporting and debugging far simpler—often saving lengthy and difficult conversations by phone and email. Naturally, one must balance the need for remote observation with the sensitivity of the environment and users in question.

7.6   SUMMARY

Pervasive display research is still in its infancy. In this chapter we have attempted to provide researchers with a starting point for projects in this area—considering a number of example research questions, some basic methodologies and sharing our experiences of conducting research in this field.

A particular problem with public displays is the questions for the validity of the data. Most studies can be criticized as not exercising sufficient control over confounding variables (internal validity), not generalizing findings to cover other settings and situations (external validity) or not testing a realistic situation (ecological validity). Internal, external and ecological validity usually cannot be achieved at the same time. Instead, studies must often sacrifice one or two of them to improve the third. In HCI, internal validity is often prioritized above the other two, leading to highly controlled lab studies with rather low ecological validity. In contrast, public displays are, by nature, a very social phenomenon. Behavior in the public space may be very different than what is predicted by lab studies [86]. Hence, for pervasive display researchers ecological validity is often prioritized above internal and external validity.

In the following chapter we provide two case studies of pervasive display research projects that have created large-scale research deployments in order to provide ecological validity.

CHAPTER 8

Case Studies

8.1 THE ROLE OF CASE STUDIES

As we saw in the last chapter, research into pervasive display systems is often conducted through the creation of prototypes that are deployed "in the wild." This provides a chance for researchers to study how users interact with such systems outside of the confines of the laboratory and provides insights into performance under realistic situations. Longitudinal studies are particularly interesting as they help mitigate against novelty effects when conducting user experiments.

In this chapter we present case studies of two research prototypes that the authors have contributed to, i.e., e-Campus and Digifieds, to help illustrate the type of work being undertaken in the field.

8.2 E-CAMPUS

8.2.1 INTRODUCTION

The e-Campus project is a long-running project at Lancaster University [182]. Begun in 2004, the project was the result of a large (£0.5m) investment intended to encourage the development of a public display research testbed on the university campus. From the outset, e-Campus was required to serve as a "laboratory" for other researchers and to be able to accommodate a range of novel applications and content types as well as meeting the needs of traditional digital signage users.

The e-Campus system consists of a variety of different installations including LCD screens, small office signs and large projector installations (Figure 8.1). At the time of writing, e-Campus

FIGURE 8.1: Examples of e-Campus deployments. (Underpass image photographer David Molyneaux, 2005)

has been fully operational for 9 years, has grown to 30 installations around campus and is in daily use as the main digital signage and emergency alert solution on our campus. As of 2011, the system had 81 individual users who had created 3,700 content items (nearly 5GB). In addition to its role as a digital signage system, e-Campus has been used for arts festivals, numerous student projects and as the basis for many research projects on interaction with public displays.

8.2.2 RESEARCH AREAS

The e-Campus project has explored a number of research areas including systems infrastructure and APIs for signage development, personalization, interactive applications and user interfaces for content creators. The work on user interfaces for content creators has already been discussed in Chapter 6, but we summarize the system software and applications work below.

Systems Software and APIs for Signage Development

The e-Campus system was designed to offer an API that enabled researchers and content authors to create sophisticated applications for presentation on e-Campus. The computational model that emerged is an abstract form of a typical hardware deployment. It consists of a small number of conceptual entities: displays, applications, schedulers and handlers. A logical display may represent a physical display, but it could also be a specific region on a screen or a particular frame buffer. In the majority of cases, applications are actually wrappers around content renderers for image, movie or web page media types. Schedulers are the back-end logic of an application, running on an application server. Handlers resolve conflicts for physical display resources between logical displays and deal with any hardware-specific issues for a display such as monitoring the maximum on-time for a projector. New schedulers are written whenever a requirement is introduced whose needs are not met by the existing set of schedulers.

The scheduling API consists of four conceptually simple operations—CreateApplication, ChangeState, Transition and TerminateApplication—that can be applied to displays. These operations are surprisingly powerful: Arguments can refer to multiple displays or groups of applications. Sequences of operations can be associated with a transactional block that can fail atomically—enabling operations that involve multiple displays or multiple content items that must be available simultaneously to either succeed or fail without affecting the visual state of the displays. The following Python code snippet illustrates the case of synchronizing the play-out of two pieces of content across two displays:

```
try:
  gid = api.MakeGroupId()
  t = transaction( api, None )
```

```
# Create renderers
(worked, pid) = t.CreateApplication('display-1',
        "http://e-content/~demo/cycling1.mpg", gid)
(worked, pid) = t.CreateApplication('display-2',
        "http://e-content/~demo/cycling2.mpg", gid)

# Cause renderers to prefetch content (note use of group id)
t.ChangeState(gid, APPLICATION_STATE_PREPARED)

# Make content visible
t.Transition(DISPLAY_ID_ALL, gid, APPLICATION_STATE_VISIBLE)
t.commit()

except transaction_aborted, msg:
  print "Can't display cycling video", msg
```

The display components take care of arbitrating requests using a simple scheme based on pre-emptable locks (an application has a display resource providing it is free or has higher priority than the application already holding it).

The computational model was implemented as a set of Python processes that communicate via a shared publish-subscribe event channel. This approach has several advantages, including the ability to inspect the system at run-time to determine its status or liveness and support for post hoc analysis after failures to determine the cause. Events can also be injected or scripted to manually orchestrate and extend the system. However, the event channel also proved the Achilles heel of the entire e-Campus system—the most common cause of outages in e-Campus related to communications problems between displays and the event channel.

The API has been used to create a number of schedulers for specific uses, including a primary scheduler for handling the placement of day-to-day digital signage content and various interaction schedulers (games, digital posters and so on) that are triggered by user interaction, including SMS messages or Bluetooth sightings.

While this API gives complete flexibility, it also became clear there was a need for a "high-level API" that was simple to use from within web-based applications (e.g., via AJAX). An HTTP-based API was created for this purpose that enables applications to create a "request" to display a "playlist" of content. A fairly powerful set of optional display constraints can be set, including the time, duration, number and gap between repetitions, displays, or presence of particular users (detected using Bluetooth MAC address). A scheduler process evaluates these constraints and the content is displayed when the appropriate conditions are met.

Applications

E-Campus is unusual in the research field in that it has been in daily use for over eight years. The principal applications that have run on the network have focused on delivering regular signage functionality. However, during its lifetime the e-Campus project has also tested a wide range of applications aimed at stimulating engagement between displays and users.

Content creation applications. To support regular signage applications, the e-Campus project developed the e-Channels system described in Chapter 6. In addition, the project team created a simple wizard that allowed users to create simple context-sensitive posters for display on the signage network. For example, users could create a poster that would be displayed when a specific user's phone (identified by the Bluetooth MAC address) was close to a display. To meet the needs of the university's administration, an application for posting emergency announcements was also developed. This enables members of the emergency response team to post announcements on the signage network that are either multiplexed with existing content or which pre-empt existing content depending on the priority assigned.

Interactive applications and personalized content. The project initially explored the use of SMS messages to support interaction with a map tailored to a display's location. Users could "text" to an advertised phone number to trigger the map and highlight a route to their destination. A downloadable J2ME application offered lower-latency interaction via Wi-Fi for compatible phones. Neither method had any lasting impact: the unpredictable latency, potential cost and effort of sending commands via SMS posed significant usability problems. The return on the effort invested by the user installing the application was not high enough, since the application was narrowly focused on the map rather than on wider interaction possibilities.

In an effort to overcome the limitations of SMS and phone applications, the project investigated Bluetooth as a possible alternative [48]. The main idea was to exploit Bluetooth-capable mobile phones, using their device name as a method for issuing commands to tailor content of nearby displays. This technique incurs no cost, there is no need to install an application and delays are bounded if not entirely consistent. The project supported commands to access Flickr photos, YouTube videos, Google keyword searches and web pages via "tinyurl" short links. While users found the system easy to use, and initial concerns about privacy or users' reluctance to change their Bluetooth names did not appear to be founded, the applications were simply not considered compelling. Despite having an extensive marketing push during a campus open-day with many thousands of visitors, the system still only managed to attract a handful of users. We note that these days many phones do not turn or keep Bluetooth on by default and in some cases (e.g., the Apple iPhone) the name cannot be changed without docking the device, substantially undermining the technique for these users.

E-Campus also used simple Bluetooth scanning to trigger content—though the time it takes to complete a Bluetooth scan makes this a far from ideal approach. Students at Lancaster developed one of the most imaginative pieces of content: a simple mixed reality game called "capture the flag." Based on the popular paint-balling game of the same name, players had to "take" an enemy stronghold (particular displays) by grouping together in front of it until their Bluetooth devices were detected. Naturally, the enemy players competed to do the same thing. The winners were the team with the most displays "captured." The game was surprisingly exciting to play, though it caused considerable consternation to other campus users as players ran around campus. This was one of the few uses of the displays that have exploited the physical separation and situated nature of the displays.

The project also conducted a series of collaborations with artists that led to the production of high-quality, non-traditional content for the signage system. Examples included specially commissioned videos and sequences of poetry and video clips that are shown when a user is detected at each display—essentially providing an experience that follows the user. While this idea was potentially very interesting, the Bluetooth-based presence detection system performed very poorly, and viewers often had to linger in front of a display to see the content, or worse, the content had already finished playing by the time the viewer could actually see the display (as opposed to simply being within Bluetooth radio range).

8.2.3   EXPERIENCES WITH CONTENT CREATION AND SHARING

The e-Channels content sharing system developed as part of e-Campus has been in continuous use since May 2008 and has proved to be highly effective in the university environment. The system has allowed display owners to adapt their display's use to their particular context and needs. Despite owners not being able to "see into" channels at the point of subscription, from the usage traces of e-Channels it can be observed that they do trust one another as content providers (actively subscribe to each others' channels). Similarly, it is content providers' responsibility to ensure that content in their channels is appropriate for the displays' audience and that, for example, copyright material is not shown—there have been almost no instances of abuses of this trust, and no obvious copyright violations.

Considerable variation is present between the different user groups and stakeholders in the system (see Figure 8.2(a)). The frequency of content generation varies dramatically (Figure 8.2(b))—for many users, the system needs to be optimized for occasional use (see [43]). The split between shared and non-shared (private) channels is almost exactly 50:50, but there are extremes: some users create primarily shared content, whereas others focus on producing private content for their own display. Channels are used effectively as organizational tools for grouping content. There is an incentive to keep reusing public channels as this avoids burdening users with the need to keep modifying their subscriptions. Some users exploited this, going directly to the file share for their channel and manipulating their content files without going via the web interface at all.

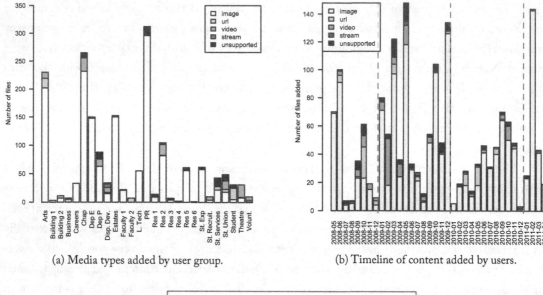

(a) Media types added by user group.　　(b) Timeline of content added by users.

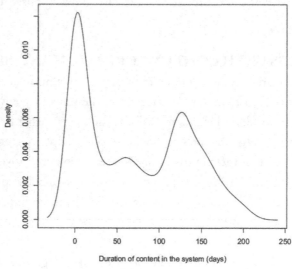

(c) Lifetime of content (probability density).

FIGURE 8.2: Media types added by user group, lifetime of content (probability density) and analysis of content lifetime, media and type.

By far the most common type of content is images (83%), with video (7%) being the next most common (see Figure 8.2(a)). Anecdotally, images are easier to produce given the expertise, time and toolchains available. The lifetime of content loosely follows the coarse timetable of the university—existing for between 1 and 3 months or more, though news and announcements of events are in the system for considerably shorter periods (Figure 8.2(c)).

Several user groups have developed a sophisticated understanding of how the system behaves and are able to subvert it to a certain extent: multiple copies of content means it will appear more often; creating long videos of static content "tricks" the scheduler into presenting it for longer—and this is arguably simpler for users and developers than offering more complex controls over content play-out. This hints that users will appropriate the system and bend it to their needs, *providing it is sufficiently intuitive and flexible*. However, such abuses, while ostensibly a positive thing where they have been used benignly to raise visibility of important content (as in our case), could also be misused in other settings.

There have been only two incidents where we have been asked to act to remove content. The first was due to "decontextualization"—a video with sound was shown during an exam, though the provider could not have known how their channel was to be used (we're certainly more careful about enabling audio in certain locations now!). The second was due to the "situated" nature of the display: a college principal was concerned that the audience would mistakenly perceive a particular anti-religious viewpoint of one content item as being endorsed by the college. Being able to *trace back a content item to its source* is an essential feature of the system in order to support handling of requests from stakeholders to remove content that is deemed inappropriate.

8.3 DIGIFIEDS

Digifieds (Figure 8.3) is a *digital public notice area* (i.e., a notice board) that has been developed by researchers at the Universities of Duisburg-Essen and Stuttgart. One of the authors of this lecture was involved in its design and implementation. Digifieds illustrates the use of a holistic approach to designing, developing and deploying a shared and networked public display application.

8.3.1 DESIGN AND DEVELOPMENT

Informing the Design

Digifieds was the result of a thorough analysis of traditional public notice areas. Prior to the development, the researchers conducted an ethnographic study in 29 locations, based on photo logs and interviews. The aim was to identify stakeholders, mechanisms for creating, posting, and exchanging content, access control and common practices around analog displays [2, 5]. The study revealed a number of interesting findings:

FIGURE 8.3:   Deployment of Digifieds in Oulu: market square (left) and public library (right).

1. With almost no exception, the content posted on public notice areas relates to the local area or community using the space (e.g., notifications of local events). Hence, the developers decided that posting procedures should support locality and provide a way to restrict content to a certain neighborhood.

2. Notice boards often reflect the different agendas of the owners of the space/display. Whereas some owners use displays as decoration, others provide them as a service for customers in an attempt to increase the importance of the space to the community and customers. As a result, Digifieds gives owners overall control over the board profile and provides features allowing them to easily choose which postings to allow or remove.

3. Notice boards often feature a wide variety of posts—ranging from handwritten notes through printed notes enhanced with images and maps to professionally designed advertisements. Digifieds was designed to facilitate this range of posts and supports ad hoc posters coincidentally approaching the display, sophisticated posters who prepare content in advance, as well as professional posters.

4. The ability to "take away" information is crucial for the success of classifieds and event promotions. As a result, traditional notices either provide pointers for viewers to retrieve information (e.g., a URL or telephone number) or a copy of the content in the form of tearaways. The developers of Digifieds have attempted to provide similar capabilities using mobile devices.

System Description and Implementation

The Difigieds platform is based on a client-server architecture. A central server component is responsible for data management and storage and supports access to content through a RESTful API. A web-based *display client* allows users to create and retrieve classified ads from a touch-enabled public display using an on-screen keyboard. Classified ads consist of a title, description, supplementary material such as images or videos and an email address for the provider of the classified ad. The display client uses AJAX to create an interactive UI capable of attracting and enticing people through immediate feedback. HTML5 and CSS are used to layout the content, and a browser in kiosk mode runs the client. Using asynchronous HTTP requests, the display client periodically polls for data changes. If there is any new content, the corresponding GUI elements are updated. The internal browser cache minimizes the data traffic and is used for media documents (images, videos, HTML, CSS). The browser's local storage API saves the classified's data in JSON format even between browser sessions or in network-loss situations.

An Android-based *mobile client* allows content to be created on-the-go (pictures and videos taken via the mobile phone camera can be included) and to be posted on an arbitrary display. Furthermore, classified ads can be retrieved from a display and be locally stored for later review. Finally, the *web client* is intended for desktop users and allows professional designs to be created (e.g., an event flyer). In contrast to the display client where content that is created using the on-screen keyboard instantly appears on the screen, content created using mobile or web clients needs to be transferred to the screen by means of different interaction techniques (see Section 8.3.2). Note that while remote posting is technically feasible, the developers of Digifieds chose to requires users to come to the display personally in order to transfer their content as a means of helping to preserve the locality of a display. In common with many commercial signage systems, Digifieds also supports the notion of display groups—allowing content to be easily posted to a number of displays at a time.

8.3.2   INTERACTION TECHNIQUES

A number of interaction techniques are provided to support the exchange of content between a display and a mobile phone (for posting or retrieving content) [5]:

Phone/display touch.   The user can touch the display with the phone at an arbitrary position. In the posting mode, the digital classified created on the phone is transferred to the display (activated) and inserted at the touched position (Figure 8.4(a)). In the retrieval mode, the digital classified located at the position where the user touches the screen with the phone is transferred to the mobile phone. The phone/display touch feature is implemented by synchronizing actions between the display and the phone. Once a touch gesture is detected, the phone and display are matched via timestamps. Subsequently, the selected post is transferred

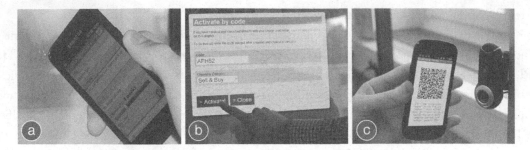

FIGURE 8.4: Posting content on a PNA—(a) phone/display touch, (b) alphanumeric code, (c) QR code on phone display or printed paper.

via the digital classifieds server between the devices. This method was primarily developed for displays with IR touch frames.

Alphanumeric code. Similar to the phone/touch feature, a five-character alphanumeric code (e.g., 4XB6A) can be used to activate a digital classified (Figure 8.4(b)). This code is associated with each post and available on the mobile phone after creating the post. It then needs to be entered on the public display of the user's choice. Furthermore, alphanumeric codes are also used for content retrieval. The code can be entered into a form on the website or directly into the mobile phone client to download the post.

QR code. Based on the alphanumeric code a QR code is also created. To enable posting, the QR code can be displayed on the mobile phone's screen and captured by the public display's camera (Figure 8.4(c)). QR codes are also displayed next to each digital classified shown on the display. These codes can be scanned using the mobile phone and either used to open the classified in the mobile browser or, if it is scanned with the Digifieds mobile client, be transferred and stored on the phone.

Technically, content created using mobile or web client is instantly transferred to the server. Using any of the above interaction techniques simply activates the classified ad for a certain display and triggers the display client to pull the content from the server.

In addition to interaction based on mobile phones, the Digifieds system also supports users without a phone via email and printouts.

Email. For each classified ad a form is provided where the user can enter their email address. The classified ad (text, images) are then sent to the user together with a link to the online version of the classified so that the user can subsequently contact the creator.

Printout.   In cases where a printer is available in the vicinity of the display, the Digifieds platform supports the printout of a classified ad. The printout includes the ad's alphanumeric code that can be entered on the website in order to contact the creator.

8.3.3   DEPLOYMENT AND EVALUATION

Digifieds was a finalist of UbiChallenge 2011 [2] and subsequently deployed in Oulu as part of their pervasive display system. The application is still available as of the time of writing. The Digifieds platform has been evaluated in a number of trials both in the lab and in the field and with regard to a number of different research questions, including usability, performance and privacy.

Investigation of Suitable Interaction Techniques

Prior to the deployment in Oulu, researchers investigated and compared Digified's different interaction techniques in a lab study. In a first step, scenarios were created that resembled realistic use cases for the Digifieds application, including (1) passing by a display and deciding to post/retrieve content using the on-screen keyboard, (2) creating content using the mobile client while being on-the-go (e.g., commuting on the train) and uploading it to a display and (3) creating content at home and exchanging the content by means of paper printouts. Concrete tasks where assigned to each scenario (e.g., *"Imagine you are at the display in a shopping center and want to sell your bike. Please create a classified ad using the display client and post it in the 'Sales' category."*).

In total, 20 participants with no CS background were recruited for the evaluation and had to perform the tasks in counterbalanced order (repeated measures design). Researchers measured task completion time and length of posts, ran usability tests (SUS questionnaire for each condition) and conducted semi-structured interviews. The results of the study showed significant differences in performance and perceived usability based on the age and technical skills of the users. Younger participants performed significantly better with the mobile phone techniques (posting and retrieving content). Tech-savvy users (i.e., users with a smartphone, that have unlimited Internet access and that make extensive use of their phones) are not only significantly faster on the mobile phone, but they also prefer the mobile phone–based techniques and write significantly more text. Detailed results can be found in Alt et al. [5].

Privacy and Content

Following up on findings from the lab, researchers focused on privacy issues as well as on content during the field trial in Finland. The field trial included observations, questionnaires and semi-structured interviews with 61 participants and an analysis of the content that was created during the first 2 months of deployment.

The questionnaires revealed no significant difference in users' privacy concerns between traditional and digital public notice areas. This is interesting, because in the digital case it is not clear for the user who has access to the data and where it appears. The questionnaires also suggest that users feel that the mobile phone can significantly help preserve their privacy, particularly with regard to sensitive information, as it makes it more difficult for bystanders to see this information being entered (i.e., shoulder surfing entry on a phone is more difficult than shoulder surfing someone entering data onto a large public display). With respect to content, users mainly expected locally relevant content, such as sales or event information.

8.4 SUMMARY

In this chapter we have presented two case studies of research prototypes that help illustrate the type of work being undertaken in the field of pervasive displays. In the remainder of this lecture we consider the future for pervasive displays and present some concluding remarks.

CHAPTER 9

Conclusion

9.1   RECAP

In this lecture we have endeavored to provide a comprehensive introduction to the field of pervasive displays. In Chapter 1 we outlined the basic characteristics of display networks that make them such an interesting topic of study. To provide a sense of the rich history of the subject, in Chapter 2 we presented a chronological view of pervasive display research. In Chapter 3 we considered what the near future might hold for display networks—describing a series of compelling applications that are being postulated for future display networks. Taken together, chapters Chapter 2 and Chapter 3 illustrated how pervasive displays are evolving from simple presentation devices to sophisticated interactive terminals providing access to highly personalized services.

Creating pervasive display systems raises a wide range of challenges and requires designers to make a series of important trade-offs, and we dedicated four chapters to key aspects of pervasive display design: audience engagement, display interaction, system software and system evaluation. In Chapter 4 we described current models that attempt to allow designers to understand how viewers and passersby might engage with displays. Once a user has become engaged, most future systems will need to support interaction with the display. In Chapter 5 we described the state-of-the-art in display interaction—focusing mainly on interaction techniques that involve a mobile device but also including a brief survey of other interaction techniques including gesture-based displays. While many different implementations are possible, important ideas relating to software that underpins pervasive display networks are described in Chapter 6.

Evaluation of systems is always important to researchers, and in Chapter 7 we described a series of evaluation techniques that can be used in the context of pervasive display research. To demonstrate some of the practical implications of the tools and techniques described, we presented in Chapter 8 two case studies: e-Campus and Digifieds.

9.2   KEY CONSIDERATIONS IN DESIGNING FUTURE PERVASIVE DISPLAY NETWORKS

Traditionally, display network designers have been concerned with issues such as the size and location of the physical display and selection of content that will attract viewers. In the future we believe the following topics will be central to the design of any display network:

Openness and control. Innovation in display networks is most likely to occur if we are able to open up future networks to a wide range of content and applications from new sources. This represents a fundamental challenge to the closed networks of today, and how to balance openness with the need to control screen access is likely to prove a significant design challenge.

Tension between personalization and privacy. It is clear that displays will have increased value if their content can be tailored automatically to the needs and interests of nearby people. However, the requirement to decide when personalized content can be presented without impacting on a viewer's privacy coupled with the need to manage the collection, storage and exchange of data to facilitate this represent fundamental challenges to the adoption of pervasive display networks.

Support for situatedness. In addition to tailoring content to specific users, considerable value can be obtained from ensuring displays evolve to provide content appropriate to their situation. Ensuring the appropriateness of displays in both fixed and dynamic situations (such as when displays themselves are attached to vehicles for example) is a key challenge.

Enticing interaction. There are currently no accepted techniques for rich user interaction with public displays. The lack of appropriate input devices, the transient nature of interactions and the potential presence of multiple users interacting with displays in a single place combine to make this an extremely challenging problem. This lack of accepted techniques limits the range of applications that can be usefully deployed and is likely to restrict the potential of pervasive display networks.

While not an exhaustive list, the considerations described above help illustrate the richness of the pervasive displays design space.

9.3   THE FUTURE

One might expect that the applications' users wish to run on pervasive public displays would have changed radically within the last 16 years. Regular desktop computers and, especially, mobile devices have seen the emergence of entirely new classes of applications that have transformed usage patterns. However, our experience is that the applications conceived for public displays have shown remarkable resilience to change.

Our first ever attempt at a public display system showed a carousel of news and departmental events that could be interrupted by personalized content when the presence of a user was detected. This personalized content was almost always a tailored news and sports feed, local weather updates and information on the user's email (typically the new message count). In recent months we have run many focus groups with viewers and owners of public display systems. It appears that despite radical change in many application areas, potential display users are still drawn to the same set of applications that were conceived over a decade ago.

It is up to the designers of future pervasive display networks to harness the opportunities presented by new technologies to innovate in this space and deliver true value to viewers. The drive toward open display networks is likely to serve as an important catalyst in this regard—enabling many more researchers and developers to create new applications. This need (and opportunity) for innovation is what makes pervasive displays such a fascinating field of research and we look forward to many exciting new applications emerging as the field develops.

Bibliography

[1] R. Adams and C. Russel. Lessons from ambient intelligence prototypes for universal access and the user experience. In C. Stephanidis and M. Pieper, editors, *Universal Access in Ambient Intelligence Environments*, volume 4397 of *Lecture Notes in Computer Science*, pages 229–243. Springer Berlin Heidelberg, 2007. DOI: 10.1007/978-3-540-71025-7. 41

[2] F. Alt, T. Kubitza, D. Bial, F. Zaidan, M. Ortel, B. Zurmaar, T. Lewen, A. Shirazi, and A. Schmidt. Digifieds: Insights into deploying digital public notice areas in the wild. In *Proceedings of the 10th International Conference on Mobile and Ubiquitous Multimedia*, MUM '11, pages 165–174, New York, NY, USA, December 2011. ACM. DOI: 10.1145/2107596.2107618. 2, 45, 54, 71, 72, 73, 74, 75, 85, 89

[3] F. Alt, J. Müller, and A. Schmidt. Advertising on public display networks. *Computer, IEEE*, 45(5):50–56, May 2012. DOI: 10.1109/MC.2012.150. 22

[4] F. Alt, A. Schmidt, and C. Evers. Mobile contextual displays. In *Proceedings of the 1st Workshop on Pervasive Advertising*, PerAd'09, 2009. 72

[5] F. Alt, A. S. Shirazi, T. Kubitza, and A. Schmidt. Interaction techniques for creating and exchanging content with public displays. In *Proceedings of the SIGCHI Conference on Human Factors in Computing Systems*, CHI '13, pages 1709–1718, New York, NY, USA, 2013. ACM. DOI: 10.1145/2470654.2466226. 47, 71, 85, 87, 89

[6] S. Antifakos and B. Schiele. Laughinglily: Using a flower as a real world information display. In *Proceedings of the 5th International Conference on Ubiquitous Computing*, Ubicomp '03, October 2003. DOI: 10.1.1.10.2743. 10

[7] Artichoke. The Telectroscope by Paul St George. 22 May–15 June 2008. London | New York. http://www.talktalk.co.uk/telectroscope/home.php [Last accessed: October 2012], 2008–2012. 8

[8] M. T. Authority. MTA to add more On the Go! touch-screen travel stations. http://new.mta.info/mta-add-more-go-touch-screen-travel-stations-0 [Last accessed: July 2013], 2013. 48

[9] P. Ball. Tv on a t-shirt. http://www.nature.com/news/1998/020520/full/news020520-4.html [Last accessed: November 2012], May 2002. 12

[10] R. Ballagas, M. Rohs, and J. G. Sheridan. Sweep and Point and Shoot: Phonecam-based Interactions for Large Public Displays. In *CHI'05 Extended Abstracts on Human Factors in Computing Systems*, CHI EA '05, pages 1200–1203, New York, NY, USA, 2005. ACM. DOI: 10.1145/1056808.1056876. 70, 71, 72, 73

[11] J. E. Bardram, T. R. Hansen, and M. Soegaard. Awaremedia: a shared interactive display supporting social, temporal, and spatial awareness in surgery. In *Proceedings of the 2006 20th Anniversary Conference on Computer Supported Cooperative Work*, CSCW '06, pages 109–118, New York, NY, USA, 2006. ACM. DOI: 10.1145/1180875.1180892. 15

[12] T. Baudel and M. Beaudouin-Lafon. Charade: remote control of objects using free-hand gestures. *Communications of the ACM*, 36(7):28–35, 1993. DOI: 10.1145/159544.159562. 49

[13] R. Beale and M. Jones. Integrating situated interaction with mobile awareness. In *Proceedings of MLEARN*, 2004. 12

[14] S. Benford, M. Flintham, A. Drozd, R. Anastasi, D. Rowland, N. Tandavanitj, M. Adams, J. Row Farr, A. Oldroyd, and J. Sutton. Uncle Roy All Around You: Implicating the city in a location-based performance. In *Proceedings of Advances in Computer Entertainment*, ACE 2004. ACM Press, 2004. 76

[15] S. Berger, R. Kjelsen, and C. Narayanaswami. Using symbiotic displays to view sensitive information in public. In *Proceedings of the 3rd International Conference on Pervasive Computing and Communications*, 2005. DOI: 10.1109/PERCOM.2005.52. 54

[16] G. Beyer, F. Alt, J. Müller, A. Schmidt, K. Isakovic, S. Klose, M. Schiewe, and I. Haulsen. Audience Behavior Around Large Interactive Cylindrical Screens. In *Proceedings of the 2011 Annual Conference on Human Factors in Computing Systems*, CHI'11, pages 1021–1030, New York, NY, USA, 2011. ACM. DOI: 10.1145/1978942.1979095. 41, 70, 73, 74

[17] G. Beyer, F. Koettner, M. Schiewe, I. Haulsen, and A. Butz. Squaring the Circle: How Framing Influences User Behavior around a Seamless Cylindrical Display. In *Proceedings of the 2013 Annual Conference on Human Factors in Computing Systems*, CHI'13, New York, NY, USA, 2013. ACM. DOI: 10.1145/2470654.2466228. 70

[18] S. A. Bly, S. R. Harrison, and S. Irwin. Media spaces: Bringing people together in a video, audio, and computing environment. *Communications of the ACM*, 36(1):28–46, January 1993. DOI: 10.1145/151233.151235. 2, 8

[19] M. A. Blythe, K. Overbeeke, and A. F. Monk. *Funology: from usability to enjoyment*, volume 3. Kluwer Academic Pub, 2004. DOI: 10.1007/1-4020-2967-5. 70

[20] M. Böhlen and M. Mateas. Office Plant #1: Intimate space and contemplative entertainment. *Leonardo*, 31(5):345–348, 1998. DOI: 10.2307/1576593. 9

[21] R. A. Bolt. "put-that-there": Voice and gesture at the graphics interface. In *Proceedings of the 7th annual conference on Computer graphics and interactive techniques*, SIGGRAPH '80, pages 262–270, New York, NY, USA, 1980. ACM. DOI: 10.1145/965105.807503. 49

[22] S. Boring, M. Altendorfer, G. Broll, O. Hilliges, and A. Butz. Shoot & copy: Phonecam-based information transfer from public displays onto mobile phones. In *Proceedings of the 4th international conference on mobile technology, applications, and systems and the 1st international symposium on Computer*

human interaction in mobile technology, Mobility '07, pages 24–31, New York, NY, USA, 2007. ACM. DOI: 10.1145/1378063.1378068. 54

[23] S. Boring, D. Baur, A. Butz, S. Gustafson, and P. Baudisch. Touch projector: Mobile interaction through video. In *Proceedings of the SIGCHI Conference on Human Factors in Computing Systems*, CHI '10, pages 2287–2296, New York, NY, USA, 2010. ACM. DOI: 10.1145/1753326.1753671. 53, 72, 73

[24] R. Borovoy. *Folk Computing: Designing Technology to Support Face-to-Face Community Building*. PhD thesis, Massachusetts Institute of Technology, 2002. 11

[25] R. Borovoy, F. Martin, M. Resnick, and B. Silverman. Groupwear: Nametags that tell about relationships. In *Conference Summary on Human Factors in Computing Systems*, CHI '98, pages 329–330, New York, NY, USA, April 1998. ACM. DOI: 10.1145/286498.286799. 11

[26] R. Borovoy, F. Martin, S. Vemuri, M. Resnick, B. Silverman, and C. Hancock. Meme tags and community mirrors: Moving from conferences to collaboration. In *Proceedings of the 1998 ACM Conference on Computer Supported Cooperative Work*, CSCW '98, pages 159–168, New York, NY, USA, 1998. ACM. DOI: 10.1145/289444.289490. 11

[27] R. Borovoy, M. McDonald, F. Martin, and M. Resnick. Things that blink: Computationally augmented name tags. *IBM Systems Journal*, 35(3-4):488–495, September 1996. DOI: 10.1147/sj.353.0488. 11

[28] A. Bragdon, R. Zeleznik, B. Williamson, T. Miller, and J. J. LaViola, Jr. Gesturebar: Improving the approachability of gesture-based interfaces. In *Proceedings of the SIGCHI Conference on Human Factors in Computing Systems*, CHI '09, pages 2269–2278, New York, NY, USA, 2009. ACM. DOI: 10.1145/1518701.1519050. 50

[29] P. B. Brandtzæg, A. Følstad, and J. Heim. Enjoyment: Lessons from Karasek. In M. A. Blythe, K. Overbeeke, A. F. Monk, and P. C. Wright, editors, *Funology*, pages 55–65, Norwell, MA, USA, 2004. Kluwer Academic Publishers. DOI: 10.1007/1-4020-2967-5_6. 41

[30] H. Brignull, S. Izadi, G. Fitzpatrick, Y. Rogers, and T. Rodden. The introduction of a shared interactive surface into a communal space. In *Proceedings of the 2004 ACM Conference on Computer Supported Cooperative Work*, CSCW '04, pages 49–58, New York, NY, USA, 2004. ACM. DOI: 10.1145/1031607.1031616. 16

[31] H. Brignull and Y. Rogers. Enticing people to interact with large public displays in public spaces. In *Proceedings of the IFIP International Conference on Human-Computer Interaction*, INTERACT '03, pages 17–24. IOS Press, September 2003. DOI: 10.1.1.129.603. 16, 34, 35, 38, 39, 42, 70

[32] C. Cadoz. Le geste canal de communication homme/machine: la communication "instrumentale". *TSI. Technique et science informatiques*, 13(1):31–61, 1994. 48

[33] J. Cardoso and R. José. PuReWidgets: a programming toolkit for interactive public display applications. In *Proceedings of the 4th ACM SIGCHI symposium on Engineering interactive computing systems*, EICS '12, pages 51–60, New York, NY, USA, 2012. ACM. DOI: 10.1145/2305484.2305496. 65

[34] M. S. T. Carpendale, D. J. Cowperthwaite, and F. D. Fracchia. 3-dimensional pliable surfaces: For the effective presentation of visual information. In *Proceedings of the 8th Annual ACM Symposium on User Interface Software and Technology*, UIST '95, pages 217–226, New York, NY, USA, 1995. ACM. DOI: 10.1145/215585.215978. 14

[35] A. Chandler, J. Finney, C. Lewis, and A. Dix. Toward emergent technology for blended public displays. In *Proceedings of the 11th International Conference on Ubiquitous Computing*, Ubicomp '09, pages 101–104, New York, NY, USA, 2009. ACM. DOI: 10.1145/1620545.1620562. 4

[36] K. Cheverst, N. Davies, K. Mitchell, and A. Friday. Experiences of developing and deploying a context-aware tourist guide: The GUIDE project. In *Proceedings of the 6th Annual International Conference on Mobile Computing and Networking*, MobiCom 2000, 2000. DOI: 10.1145/345910.345916. 76

[37] K. Cheverst, D. Fitton, A. Dix, and M. Rouncefield. Exploring situated interaction with ubiquitous office door displays. In *Public, Community and Situated Displays (Workshop at CSCW 2002)*, New Orleans, LA, USA, 2002. 12

[38] K. Cheverst, D. Fitton, and A. J. Dix. Exploring the Evolution of Office Door Displays. In K. O'Hara, E. Perry, E. Churchill, and D. M. Russel, editors, *Public and Situated Displays—Social and Interactional Aspects of Shared Display Technologies*, pages 141–169. Kluwer, Dordrecht, 2003. DOI: 10.1007/978-94-017-2813-3_6. 12, 25, 74

[39] K. Cheverst, F. Taher, M. Fisher, D. Fitton, and N. Taylor. The design, deployment and evaluation of situated display-based systems to support coordination and community. In A. Krüger and T. Kuflik, editors, *Ubiquitous Display Environments*, Cognitive Technologies, pages 105–124. Springer-Verlag, Berlin, Heidelberg, 2012. DOI: 10.1007/978-3-642-27663-7_7. 12, 18

[40] K. Cheverst, N. Taylor, M. Rouncefield, A. Galani, and C. Kray. The Challenge of Evaluating Situated Display-based Technology Interventions Designed to Foster a Sense of Community. In *Proceedings of the Workshop on Ubiquitous Systems Evaluation*, USE'08, 2008. DOI: 10.1.1.142.7222. 18, 71, 73, 74

[41] A. Chew, V. Leclerc, S. Sadi, A. Tang, and H. Ishii. SPARKS. In *Extended Abstracts on Human Factors in Computing Systems*, CHI '05, pages 1276–1279, New York, NY, USA, 2005. ACM. DOI: 10.1145/1056808.1056895. 16

[42] E. F. Churchill, L. Nelson, L. Denoue, and A. Girgensohn. The plasma poster network: Posting multimedia content in public places. In *Proceedings of the 9th IFIP TC13 International Conference on Human-Computer Interaction*, INTERACT '03. IOS Press, September 2003. 14

[43] S. Clinch, N. Davies, A. Friday, and C. Efstratiou. Reflections on the long-term use of an experimental public display system. In *Proceedings of the 13th International Conference on Ubiquitous Computing*, Ubicomp '11, pages 133–142, 2011. DOI: 10.1145/2030112.2030132. 60, 83

[44] S. Clinch, N. Davies, T. Kubitza, and A. Schmidt. Designing application stores for public display networks. In *Proceedings of the 2012 International Symposium on Pervasive Displays*, PerDis '12, New York, NY, USA, 2012. ACM. DOI: 10.1145/2307798.2307808. 62

[45] S. Clinch, J. Harkes, A. Friday, N. Davies, and M. Satyanarayanan. How close is close enough? understanding the role of cloudlets in supporting display appropriation by mobile users. In *Proceedings of the 2012 IEEE International Conference on Pervasive Computing and Communication*, PerCom '12, pages 122–127, 2012. DOI: 10.1109/PerCom.2012.6199858. 53

[46] C. Cochrane, L. Meunier, F. M. Kelly, and V. Koncar. Flexible displays for smart clothing: Part I—overview. *Indian Journal of Fibre and Textile Research*, 36:422–428, December 2011. 12

[47] M. Collomb, M. Hascoët, P. Baudisch, and B. Lee. Improving Drag-and-Drop on Wall-size Displays. In *Proceedings of Graphics Interface 2005*, GI'05, pages 25–32, Waterloo, ON, Canada, 2005. Canadian Human-Computer Communications Society. 71, 72

[48] N. Davies, A. Friday, P. Newman, S. Rutlidge, and O. Storz. Using bluetooth device names to support interaction in smart environments. In *Proceedings of the 7th International Conference on Mobile Systems, Applications, and Services*, MobiSys '09, 2009. DOI: 10.1145/1555816.1555832. 26, 51, 82

[49] N. Davies, M. Langheinrich, R. José, and A. Schmidt. Open display networks: A communications medium for the 21st century. *Computer, IEEE*, 45(5):58–64, May 2012. DOI: 10.1109/MC.2012.114. 28, 29, 30

[50] A. K. Dey. *Providing Architectural Support for Building Context-Aware Applications*. PhD thesis, College of Computing, Georgia Institute of Technology, 2000. 24

[51] P. Dietz and D. Leigh. Diamondtouch: a multi-user touch technology. In *Proceedings of the 14th annual ACM symposium on User interface software and technology*, UIST '01, pages 219–226, New York, NY, USA, 2001. ACM. DOI: 10.1145/502348.502389. 47

[52] Digital Signage Networks India Pvt. Ltd. DSN—digital signage networks. http://app.dsnglobal .com/ [Last accessed: September 2012], 2012. 17

[53] D. Easterly. Bio-Fi: Inverse biotelemetry projects. In *Proceedings of the 12th Annual ACM International Conference on Multimedia*, MULTIMEDIA '04, pages 182–183, New York, NY, USA, October 2004. ACM. DOI: 10.1145/1027527.1027568. 10

[54] J. Enns, E. Austen, V. Di Lollo, R. Rauschenberger, and S. Yantis. New Objects Dominate Luminance Transients in Setting Attentional Priority. *Journal of Experimental Psychology: Human Perception and Performance*, 27(6):1287, 2001. DOI: 10.1037/0096-1523.27.6.1287. 37

[55] A. Erbad, M. Blackstock, A. Friday, R. Lea, and J. Al-Muhtadi. Magic broker: A middleware toolkit for interactive public displays. In *Proceedings of the 2008 Sixth Annual IEEE International Conference on Pervasive Computing and Communications*, PerCom '08, pages 509–514, Washington, DC, USA, 2008. IEEE Computer Society. DOI: 10.1109/PERCOM.2008.109. 52

[56] J. Falk and S. Björk. The BubbleBadge: A wearable public display. In *Extended Abstracts on Human Factors in Computing Systems*, CHI '99, pages 318–319, New York, NY, USA, May 1999. ACM. DOI: 10.1145/632716.632909. 11

[57] S. D. Farnham, J. F. McCarthy, Y. Patel, S. Ahuja, D. Norman, W. R. Hazlewood, and J. Lind. Measuring the impact of third place attachment on the adoption of a place-based community technology. In *Proceedings of the SIGCHI Conference on Human Factors in Computing Systems*, pages 2153–2156, 2009. DOI: 10.1145/1518701.1519028. 16

[58] A. M. Fass. MessyBoard: Lowering the cost of communication and making it more enjoyable. In *Doctorial Symposium at the Seventeenth Annual ACM Symposium on User Interface Software and Technology (UIST 2004)*, 2004. 14

[59] A. M. Fass, J. Forlizzi, and R. Pausch. MessyDesk and MessyBoard: Two designs inspired by the goal of improving human memory. In *Proceedings of the 4th Conference on Designing Interactive Systems: Processes, Practices, Methods, and Techniques*, DIS '02, pages 303–311, New York, NY, USA, 2002. ACM. DOI: 10.1145/778712.778754. 14

[60] A. Fatah gen Schieck, V. Kostakos, and A. Penn. The Urban Screen as a Socialising Platform: Exploring the Role of Place Within the Urban Space. In F. Eckhardt, J. Geelhaar, L. Colini, K. S. Willis, K. Chorianopoulos, and R. Henning, editors, *MEDIACITY—Situations, Practices and Encounters*, pages 285–305. Frank & Timme GmbH, 2008. 70

[61] A. Ferscha and S. Vogl. Wearable displays for everyone! *Pervasive Computing, IEEE*, 9(1):7–10, January–March 2010. DOI: 10.1109/MPRV.2010.13. 11

[62] J. Finney, S. Wade, N. Davies, and A. Friday. Flump: The FLexible Ubiquitous Monitor Project. In *Cabernet Radicals Workshop*, May 1996. 8

[63] R. S. Fish, R. E. Kraut, and B. L. Chalfonte. The VideoWindow system in informal communication. In *Proceedings of the 1990 ACM Conference on Computer-Supported Cooperative Work*, CSCW '90, pages 1–11, New York, NY, USA, 1990. ACM. DOI: 10.1145/99332.99335. 8

[64] D. Fitton, K. Cheverst, J. Finney, and A. Dix. Supporting interaction with office door displays. In *Workshop on Multi-User and Ubiquitous User Interfaces 2004*, MU3I 2004, 2004. DOI: 10.1.1.164.2417. 12

[65] S. Franconeri and D. Simons. Moving and Looming Stimuli Capture Attention. *Attention, Perception, & Psychophysics*, 65(7):999–1010, 2003. DOI: 10.3758/BF03194829. 37, 38

[66] D. Freeman, H. Benko, M. R. Morris, and D. Wigdor. Shadowguides: visualizations for in-situ learning of multi-touch and whole-hand gestures. In *Proceedings of the ACM International Conference*

on Interactive Tabletops and Surfaces, ITS '09, pages 165–172, New York, NY, USA, 2009. ACM. DOI: 10.1145/1731903.1731935. 49, 50

[67] A. Friday, N. Davies, and C. Efstratiou. Reflections on long-term experiments with public displays. *Computer, IEEE*, 45(5):34–41, May 2012. DOI: 10.1109/MC.2012.155. 18, 76

[68] T. Fuhrmann, M. Klein, and M. Odendahl. The BlueWand as interface for ubiquitous and wearable computing environments. In *Proceedings of the 5th European Personal Mobile Communications Conference*, pages 91–95. IET, April 2003. DOI: 10.1049/cp:20030225. 50

[69] K. Galloway and S. Rabinowitz. Hole-In-Space. http://www.ecafe.com/getty/HIS/ [Last accessed: October 2012], 1980. 7

[70] G. O. Goodman and M. J. Abel. Collaboration research in SCL. In *Proceedings of the 1986 ACM Conference on Computer-supported Cooperative Work*, CSCW '86, pages 246–251, New York, NY, USA, 1986. ACM. DOI: 10.1145/637069.637099. 8

[71] P. Gould. Textiles gain intelligence. *Materials Today*, 6:38–43, 2003. DOI: 10.1016/S1369-7021(03)01028-9. 12

[72] A. Grasso, M. Muehlenbrock, F. Roull, and D. Snowdon. Supporting communities of practice with large screen displays. In K. O'Hara, M. Perry, E. Churchill, and D. M. Russel, editors, *Public and Situated Displays—Social and Interactional Aspects of Shared Display Technologies*, pages 261–282. Kluwer, 2003. DOI: 10.1007/978-94-017-2813-3_11. 14

[73] S. Greenberg. Designing computers as public artifacts. In *International Journal of Design Computing: Special Issue on Design Computing on the Net (DCNet '99)*, 1999. 13

[74] E. T. Hall. *The Hidden Dimension*. Anchor Books, 1966. DOI: 10.2307/1572461. 35

[75] J. Hardy, E. Rukzio, and N. Davies. Real world responses to interactive gesture based public displays. In *Proceedings of the 10th International Conference on Mobile and Ubiquitous Multimedia*, MUM '11, pages 33–39, New York, NY, USA, December 2011. ACM. DOI: 10.1145/2107596.2107600. 3

[76] R. Hardy, E. Rukzio, M. Wagner, and M. Paolucci. Exploring expressive nfc-based mobile phone interaction with large dynamic displays. In *Proceedings of the 2009 First International Workshop on Near Field Communication*, NFC '09, pages 36–41, Washington, DC, USA, 2009. IEEE Computer Society. DOI: 10.1109/NFC.2009.10. 53

[77] C. Harrison, B. Y. Lim, A. Shick, and S. E. Hudson. Where to locate wearable displays?: Reaction time performance of visual alerts from tip to toe. In *Proceedings of the SIGCHI Conference on Human Factors in Computing Systems*, CHI '09, pages 941–944, New York, NY, USA, 2009. ACM. DOI: 10.1145/1518701.1518845. 11

[78] W. R. Hazlewood, N. Dalton, P. Marshall, Y. Rogers, and S. Hertrich. Bricolage and consultation: Addressing new design challenges when building large-scale installations. In *Proceedings of the 8th ACM Conference on Designing Interactive Systems*, DIS '10, pages 380–389, New York, NY, USA, 2010. ACM. DOI: 10.1145/1858171.1858244. 10, 19

[79] T. Heikkinen, T. Lindén, T. Ojala, H. Kukka, M. Jurmu, and S. Hosio. Lessons learned from the deployment and maintenance of ubi-hotspots. In *Proceedings of the 4th International Conference on Multimedia and Ubiquitous Engineering*, MUE '10, August 2010. DOI: 10.1109/MUE.2010.5575054. 17, 18

[80] J. M. Heiner, S. E. Hudson, and K. Tanaka. The information percolator: Ambient information display in a decorative object. In *Proceedings of the 12th Annual ACM Symposium on User Interface Software and Technology*, UIST '99, pages 141–148, New York, NY, USA, 1999. ACM. DOI: 10.1145/320719.322595. 10

[81] Helsinki Institute for Information Technology. Citywall. http://citywall.org/ [Last accessed: February 2013], 2013. 18

[82] O. Hilliges, S. Izadi, A. D. Wilson, S. Hodges, A. Garcia-Mendoza, and A. Butz. Interactions in the Air: Adding Further Depth to Interactive Tabletops. In *Proceedings of the 22nd Annual ACM Symposium on User Interface Software and Technology*, UIST'09, pages 139–148, New York, NY, USA, 2009. ACM. DOI: 10.1145/1622176.1622203. 39

[83] U. Hinrichs and S. Carpendale. Gestures in the Wild: Studying Multi-Touch Gesture Sequences on Interactive Tabletop Exhibits. In *Proceedings of the 2011 Annual Conference on Human Factors in Computing Systems*, CHI'11, pages 3023–3032, New York, NY, USA, 2011. ACM. DOI: 10.1145/1978942.1979391. 39

[84] D. Holstius, J. Kembel, A. Hurst, P.-H. Wan, and J. Forlizzi. Infotropism: Living and robotic plants as interactive displays. In *Proceedings of the 5th Conference on Designing Interactive Systems: Processes, Practices, Methods, and Techniques*, DIS '04, pages 215–221, New York, NY, USA, July 2004. ACM. DOI: 10.1145/1013115.1013145. 10

[85] S. Houde, R. Bellamy, and L. Leahy. In search of design principles for tools and practices to support communication within a learning community. *SIGCHI Bulletin*, 30(2):113–118, April 1998. DOI: 10.1145/279044.279171. 2, 13

[86] E. M. Huang, A. Koster, and J. Borchers. Overcoming assumptions and uncovering practices: When does the public really look at public displays? In *Proceedings of the 6th International Conference on Pervasive Computing*, Pervasive '08, pages 228–243, Berlin, Heidelberg, May 2008. Springer-Verlag. DOI: 10.1007/978-3-540-79576-6_14. 3, 22, 37, 46, 71, 73, 77

[87] E. M. Huang and E. D. Mynatt. Semi-public displays for small, co-located groups. In *Proceedings of the SIGCHI Conference on Human Factors in Computing Systems*, CHI '03, pages 49–56, New York, NY, USA, 2003. ACM. DOI: 10.1145/642611.642622. 2, 15

[88] H. Hutchinson, W. Mackay, B. Westerlund, B. B. Bederson, A. Druin, C. Plaisant, M. Beaudouin-Lafon, S. Conversy, H. Evans, H. Hansen, N. Roussel, and B. Eiderbäck. Technology Probes: Inspiring Design for and with Families. In *Proceedings of the SIGCHI Conference on*

Human Factors in Computing Systems, CHI'03, pages 17–24, New York, NY, USA, 2003. ACM. DOI: 10.1145/642611.642616. 74

[89] INFOSCREEN Austria, Gesellschaft für Stadtinformationsanlagen GmbH. Infoscreen—your city channel | infoscreen. http://www.infoscreen.at/www/homepage.php?lng=EN [Last accessed: January 2013], 2005-2009. 17

[90] INFOSCREEN NETWORKS PLC. Infoscreen networks. http://www.infoscreennetworks.com/ [Last accessed: February 2013], 2011-2012. 17

[91] S. S. Intille. Change Blind Information Display for Ubiquitous Computing Environments. In *Proceedings of the 4th international Conference on Ubiquitous Computing*, UbiComp'02, pages 91–106, London, UK, UK, 2002. Springer-Verlag. DOI: 10.1007/3-540-45809-3_7. 38

[92] L. Itti and P. Baldi. Bayesian Surprise Attracts Human Attention. *Vision Research*, 49(10): 1295–1306, 2009. DOI: 10.1016/j.visres.2008.09.007. 38

[93] S. Iyengar and M. Lepper. Rethinking the Value of Choice: A Cultural Perspective on Intrinsic Motivation. *Journal of Personality and Social Psychology*, 76(3):349, 1999. DOI: 10.1037/0022-3514.76.3.349. 41

[94] S. Izadi, H. Brignull, T. Rodden, Y. Rogers, and M. Underwood. Dynamo: A public interactive surface supporting the cooperative sharing and exchange of media. In *Proceedings of the 16th Annual ACM Symposium on User Interface Software and Technology*, UIST '03, pages 159–168, New York, NY, USA, 2003. ACM. DOI: 10.1145/964696.964714. 16, 59

[95] G. Jacucci, A. Morrison, G. T. Richard, J. Kleimola, P. Peltonen, L. Parisi, and T. Laitinen. Worlds of Information: Designing for Engagement at a Public Multi-Touch Display. In *Proceedings of the 28th international Conference on Human Factors in Computing Systems*, CHI'10, pages 2267–2276, New York, NY, USA, 2010. ACM. DOI: 10.1145/1753326.1753669. 18, 39, 72, 74

[96] N. Jafarinaimi, J. Forlizzi, A. Hurst, and J. Zimmerman. Breakaway: An ambient display designed to change human behavior. In *Extended Abstracts on Human Factors in Computing Systems*, CHI '05, pages 1945–1948, New York, NY, USA, April 2005. ACM. DOI: 10.1145/1056808.1057063. 10

[97] G. Jancke, G. D. Venolia, J. Grudin, J. J. Cadiz, and A. Gupta. Linking public spaces: Technical and social issues. In *Proceedings of the SIGCHI Conference on Human Factors in Computing Systems*, CHI '01, pages 530–537, New York, NY, USA, March 2001. ACM. DOI: 10.1145/365024.365352. 8

[98] B. Jeong, J. Leigh, A. Johnson, L. Renambot, M. Brown, R. Jagodic, S. Nam, and H. Hur. Ultrascale collaborative visualization using a display-rich global cyberinfrastructure. *Computer Graphics and Applications, IEEE*, 30(3):71–83, 2010. DOI: 10.1109/MCG.2010.45. 65

[99] B. Johanson and A. Fox. The event heap: A coordination infrastructure for interactive workspaces. In *Proceedings of the Fourth IEEE Workshop on Mobile Computing Systems and Applications*, WMCSA '02, Washington, DC, USA, 2002. IEEE Computer Society. DOI: 10.1109/MCSA.2002.1017488. 77

[100] J. Jonides and S. Yantis. Uniqueness of Abrupt Visual Onset in Capturing Attention. *Attention, Perception, & Psychophysics*, 43(4):346–354, 1988. DOI: 10.3758/BF03208805. 37

[101] R. José, N. Otero, S. Izadi, and R. Harper. Instant Places: Using bluetooth for situated interaction in public displays. *Pervasive Computing, IEEE*, 7, 2008. DOI: 10.1109/MPRV.2008.74. 51, 59

[102] W. Ju and D. Sirkin. Animate Objects: How Physical Motion Encourages Public Interaction. In *Proceedings of the 5th international Conference on Persuasive Technology*, PERSUASIVE'10, pages 40–51, Berlin, Heidelberg, 2010. Springer-Verlag. DOI: 10.1007/978-3-642-13226-1_6. 37, 39

[103] K. Karahalios and J. Donath. Telemurals: Linking remote spaces with social catalysts. In *Proceedings of the SIGCHI Conference on Human Factors in Computing Systems*, CHI '04, pages 615–622, New York, NY, USA, April 2004. ACM. DOI: 10.1145/985692.985770. 8

[104] A. Khan, J. Matejka, G. Fitzmaurice, and G. Kurtenbach. Spotlight: Directing Users' Attention on Large Displays. In *Proceedings of the SIGCHI Conference on Human Factors in Computing Systems*, CHI'05, pages 791–798, New York, NY, USA, 2005. ACM. DOI: 10.1145/1054972.1055082. 72

[105] C. Kray, A. Galani, and M. Rohs. Facilitating opportunistic interaction with ambient displays. In *Workshop on Designing and Evaluating Mobile Phone-Based Interaction with Public Displays at CHI 2008*, 2008. 18

[106] C. Kray, G. Kortuem, and A. Krüger. Adaptive navigation support with public displays. In *Proceedings of the 10th international conference on Intelligent user interfaces*, IUI '05, pages 326–328, New York, NY, USA, 2005. ACM. DOI: 10.1145/1040830.1040916. 25

[107] M. W. Krueger. *Artificial Reality II*. Addison-Wesley, 1991. 39

[108] Davies, N., M. Langheinrich, S. Clinch, I. Elhart, A. Friday, T. Kubitza, and B. Surajbali, "Personalisation and Privacy in Future Pervasive Display Networks." In: CHI '14: Proceedings of the 32th SIGCHI Conference on Human Factors in Computing Systems. Toronto, ON, Canada, April 26–May 1, 2014. New York, NY, USA: ACM, April 2014. 29, 51, 52

[109] B. Kules, H. Kang, C. Plaisant, A. Rose, and B. Shneiderman. Immediate Usability: A Case Study of Public Access Design for a Community Photo Library. *Interacting with Computers*, 16(6):1171–1193, 2004. DOI: 10.1016/j.intcom.2004.07.005. 39, 40

[110] S. Kuribayashi and A. Wakita. PlantDisplay: Turning houseplants into ambient display. In *Proceedings of the International Conference on Advances in Computer Entertainment Technology*, ACE '06, New York, NY, USA, 2006. ACM. DOI: 10.1145/1178823.1178871. 10

[111] G. Kurtenbach and E. A. Hulteen. Gestures in human-computer communication. *The art of human-computer interface design*, pages 309–317, 1990. 48

[112] J. Lazar, J. Feng, and H. Hochheiser. *Research Methods in Human-Computer Interaction*. John Wiley & Sons Inc, 2009. 72

[113] J. Y. Lee, M. S. Kim, D. W. Seo, C.-W. Lee, J. S. Kim, and S. M. Lee. Dual interactions between multi-display and smartphone for collaborative design and sharing. In *Proceedings of IEEE Virtual Reality (VR) Conference 2011*, pages 221–222, 2011. DOI: 10.1109/VR.2011.5759478. 54

[114] J. Y. Lee, M. S. Kim, D. W. Seo, C.-W. Lee, J. S. Kim, and S. M. Lee. Smart and space-aware interactions using smartphones in a shared space. In *Proceedings of the 14th international conference on Human- computer interaction with mobile devices and services companion*, MobileHCI '12, pages 53–58, New York, NY, USA, 2012. ACM. DOI: 10.1145/2371664.2371676. 54

[115] Litfest and Flax Book. Finding a language [video series]. http://www.youtube.com/watch?v=cjvfP1uxVXg&list=PL2E3F808CF9BE133B [Last accessed July 2013], 2008. 26

[116] P. Ljungstrand, S. Bjork, and J. Falk. The wearboy: a platform for low-cost public wearable devices. In *Proceedings of the Third International Symposium on Wearable Computers.*, pages 195–196, October 1999. DOI: 10.1109/ISWC.1999.806926. 11

[117] T. Malone. Toward a Theory of Intrinsically Motivating Instruction. *Cognitive Science*, 5(4):333–369, 1981. DOI: 10.1207/s15516709cog0504_2. 41

[118] T. Malone and M. Lepper. Making Learning Fun: A Taxonomy of Intrinsic Motivations for Learning. *Aptitude Learning and Instruction*, 3(3):223–253, 1987. 40

[119] J. Mankoff and B. N. Schilit. Supporting knowledge workers beyond the desktop with palplates. In *Proceedings of the SIGCHI Conference on Human Factors in Computing Systems*, CHI '97, pages 550–551. ACM, March 1997. DOI: 10.1145/258549.259030. 12

[120] P. Marshall, R. Morris, Y. Rogers, S. Kreitmayer, and M. Davies. Rethinking 'Multi-User': An In-the-Wild Study of How Groups Approach a Walk-Up-and-Use Tabletop Interface. In *Proceedings of the 2011 annual Conference on Human Factors in Computing Systems*, CHI'11, pages 3033–3042, New York, NY, USA, 2011. ACM. DOI: 10.1145/1978942.1979392. 39, 40, 42

[121] M. May, S. Scheider, R. Rösler, D. Schulz, and D. Hecker. Pedestrian flow prediction in extensive road networks using biased observational data. In *Proceedings of the 16th ACM SIGSPATIAL International Conference on Advances in Geographic Information Systems*, GIS '08, pages 67:1–67:4, New York, NY, USA, 2008. ACM. DOI: 10.1145/1463434.1463512. 69

[122] J. F. McCarthy. Using public displays to create conversation opportunities. In *Public, Community and Situated Displays (Workshop at CSCW 2002)*, New Orleans, LA, USA, 2002. 15, 16

[123] J. F. McCarthy. Promoting a sense of community with ubiquitous peripheral displays. In K. O'Hara, M. Perry, E. Churchill, and D. M. Russel, editors, *Public and Situated Displays—Social and Interactional Aspects of Shared Display Technologies*, pages 283–308. Kluwer, 2003. DOI: 10.1007/978-94-017-2813-3_12. 15

[124] J. F. McCarthy, T. J. Costa, and E. S. Liongosari. Unicast, outcast & groupcast: Three steps toward ubiquitous, peripheral displays. In *Proceedings of the 3rd International Conference on Ubiquitous*

Computing, Ubicomp '01, pages 332–345, London, UK, 2001. Springer-Verlag. DOI: 10.1007/3-540-45427-6_28. 15

[125] J. F. McCarthy, D. W. McDonald, S. Soroczak, D. H. Nguyen, and A. M. Rashid. Augmenting the social space of an academic conference. In *Proceedings of the 2004 ACM Conference on Computer Supported Cooperative Work*, CSCW '04, pages 39–48. ACM, New York, NY, USA, 2004. DOI: 10.1145/1031607.1031615. 16

[126] N. Memarovic, I. Elhart, and M. Langheinrich. FunSquare: First experiences with autopoiesic content. In *Proceedings of the 10th International Conference on Mobile and Ubiquitous Multimedia*, MUM '11, pages 175–184, New York, NY, USA, 2011. ACM. DOI: 10.1145/2107596.2107619. 60, 71, 74

[127] N. Memarovic, M. Langheinrich, F. Alt, I. Elhart, S. Hosio, and E. Rubegni. Using public displays to stimulate passive engagement, active engagement, and discovery in public spaces. In *Proceedings of the 4th Media Architecture Biennale Conference: Participation*, MAB '12, pages 55–64, New York, NY, USA, 2012. ACM. DOI: 10.1145/2421076.2421086. 35

[128] A. Meschtscherjakov, W. Reitberger, T. Mirlacher, H. Huber, and M. Tscheligi. AmIQuin—An Ambient Mannequin for the Shopping Environment. In *Proceedings of the European Conference on Ambient Intelligence*, AmI'09, pages 206–214. Springer-Verlag, Berlin, Heidelberg, 2009. DOI: 10.1007/978-3-642-05408-2_25. 71

[129] D. Michelis. *Interaktive Großbildschirme im öffentlichen Raum: Nutzungsmotive und Gestaltungsregeln.* Gabler, 2009. DOI: 10.1007/978-3-8349-9451-6. 38, 41

[130] D. Michelis and M. Meckel. Why do we want to interact with electronic billboards in public space? In *Proceedings of the 2009 Workshop on Pervasive Advertising*, Pervasive '09, 2009. 39

[131] D. Michelis and J. Müller. The audience funnel: Observations of gesture based interaction with multiple large displays in a city center. *International Journal of Human-Computer Interaction*, 27(6):562–579, 2011. DOI: 10.1080/10447318.2011.555299. 34, 36, 39, 42

[132] Microsoft. Microsoft pixelsense. http://www.microsoft.com/en-us/pixelsense/default.aspx [Last accessed: July 2013], 2012. 48

[133] A. Morrison, G. Jacucci, and P. Peltonen. Citywall: Limitations of a multi-touch environment. In *Proceedings of the 2008 Workshop on Designing Multi-Touch Interaction Techniques for Coupled Public and Private Displays (at AVI 2008).*, PPD '08, 2008. 18

[134] J. Müller, F. Alt, D. Michelis, and A. Schmidt. Requirements and Design Space for Interactive Public Displays. In *Proceedings of the International Conference on Multimedia*, MM'10, pages 1285–1294, New York, NY, USA, 2010. ACM. DOI: 10.1145/1873951.1874203. 36, 37

[135] J. Müller, J. Exeler, M. Buzeck, and A. Krüger. Reflectivesigns: Digital signs that adapt to audience attention. In *Proceedings of the 7th International Conference on Pervasive Computing*, Pervasive '09,

pages 17–24, Berlin, Heidelberg, May 2009. Springer-Verlag. DOI: 10.1007/978-3-642-01516-8_3. 18

[136] J. Müller, O. Paczkowski, and A. Krüger. Situated public news and reminder displays. In *Proceedings of the European Conference on Ambient Intelligence*, AmI '07, pages 248–265, Berlin, Heidelberg, 2007. Springer-Verlag. DOI: 10.1007/978-3-540-76652-0_15. 18

[137] J. Müller, R. Walter, G. Bailly, M. Nischt, and F. Alt. Looking Glass: A Field Study on Noticing Interactivity of a Shop Window. In *Proceedings of the 2012 ACM Conference on Human Factors in Computing Systems*, CHI'12, pages 297–306, New York, NY, USA, 2012. ACM. DOI: 10.1145/2207676.2207718. 40, 51, 70, 71, 72, 73, 74, 75, 76

[138] J. Müller, D. Wilmsmann, J. Exeler, M. Buzeck, A. Schmidt, T. Jay, and A. Krüger. Display blindness: The effect of expectations on attention towards digital signage. In *Proceedings of the 7th International Conference on Pervasive Computing*, Pervasive '09, pages 1–8, Berlin, Heidelberg, May 2009. Springer-Verlag. DOI: 10.1007/978-3-642-01516-8_1. 3, 18, 71, 72, 74, 76

[139] D. Nguyen, J. Tullio, T. Drewes, and E. Mynatt. Dynamic door displays. GVU Technical Report GIT-GVU-00-30, Georgia Institute of Technology, GVU, 2000. 12

[140] Nintendo. Wii official site at nintendo. http://www.nintendo.com/wii [Last accessed: January 2013], 2013. 17

[141] D. A. Norman. Natural user interfaces are not natural. *interactions*, 17(3):6–10, May 2010. DOI: 10.1145/1744161.1744163. 49

[142] Q. U. of Technology. Welcome to nnub. http://nnub.net/ [Last accessed: January 2013], 2013. 18

[143] K. O'Hara, M. Perry, and S. Lewis. Situated Web signs and the ordering of social action. In K. O'Hara, M. Perry, E. Churchill, and D. M. Russel, editors, *Public and Situated Displays—Social and Interactional Aspects of Shared Display Technologies*, pages 105–140. Kluwer, 2003. 12

[144] T. Ojala, V. Kostakos, and H. Kukka. It's a Jungle Out There: Fantasy and Reality of Evaluating Public Displays in the Wild. *Proceedings of the First Workshop on Large Displays in Urban Life*, 4:1–4, 2011. 39, 76

[145] T. Ojala, H. Kukka, T. Lindén, T. Heikkinen, M. Jurmu, S. Hosio, and F. Kruger. UBI-Hotspot 1.0: Large-Scale Long-Term Deployment of Interactive Public Displays in a City Center. In *Proceedings of the 2010 Fifth International Conference on Internet and Web Applications and Services*, ICIW'10, pages 285–294, Washington, DC, USA, 2010. IEEE Computer Society. DOI: 10.1109/ICIW.2010.49. 17, 74

[146] Outdoor Advertising Association of America, Inc. Outdoor advertising association of america. http://www.oaaa.org/ [Last accessed: July 2013], 2013. 21

[147] B. Paras and J. Bizzocchi. Game, motivation, and effective learning: An integrated model for educational game design. In *DiGRA 2005*, 2005. 41

[148] P. Peltonen, E. Kurvinen, A. Salovaara, G. Jacucci, T. Ilmonen, J. Evans, A. Oulasvirta, and P. Saarikko. It's Mine, Don't Touch!: Interactions at a Large Multi-Touch Display in a City Centre. In *Proceeding of the 26th Annual SIGCHI Conference on Human Factors in Computing Systems*, CHI'08, pages 1285–1294, New York, NY, USA, 2008. ACM. DOI: 10.1145/1357054.1357255. 18, 38, 39, 42, 71, 72

[149] P. Peltonen, A. Salovaara, G. Jacucci, T. Ilmonen, C. Ardito, P. Saarikko, and V. Batra. Extending large-scale event participation with user-created mobile media on a public display. In *Proceedings of the 6th International Conference on Mobile and Ubiquitous Multimedia*, MUM '07, pages 131–138, New York, NY, USA, 2007. ACM. DOI: 10.1145/1329469.1329487. 18

[150] T. Pering, R. Ballagas, and R. Want. Spontaneous marriages of mobile devices and interactive spaces. *Communications of the ACM*, 48(9):53–59, 2005. DOI: 10.1145/1081992.1082020. 53

[151] T. Prante, C. Röcker, N. Streitz, R. Stenzel, C. Magerkurth, D. van Alphen, and D. Plewe. Hello.Wall—Beyond Ambient Displays. In *Adjunct Proceedings of the 5th International Conference on Ubiquitous Computing*, Ubicmop'03, pages 277–278, 2003. 34

[152] I. K. Rakkolainen, and A. K. Lugmayr. Immaterial display for interactive advertisements. In *Proceedings of the International Conference on Advances in Computer Entertainment Technology*, ACE '07, pages 95–98, New York, NY, USA, 2007. ACM. DOI: 10.1145/1255047.1255066. 10

[153] I. K. Rakkolainen, S. DiVerdi, A. Olwal, N. Candussi, T. Höllerer, M. Laitinen, M. Piirto, and K. Palovuori. The interactive FogScreen. In *ACM SIGGRAPH 2005 Emerging Technologies*, SIGGRAPH '05, New York, NY, USA, 2005. ACM. DOI: 10.1145/1187297.1187306. 10

[154] I. K. Rakkolainen, T. Erdem, Çiğdem Erdem, M. Özkan, and M. Laitinen. Interactive "immaterial" screen for performing arts. In *Proceedings of the 14th Annual ACM International Conference on Multimedia*, MULTIMEDIA '06, pages 185–188, New York, NY, USA, 2006. ACM. DOI: 10.1145/1180639.1180692. 10

[155] I. K. Rakkolainen and K. Palovuori. WAVE—a walk-thru virtual environment. In *CD Proceedings of the 6th Immersive Projection Technology Symposium*, In association with IEEE VR 2002, March 2002. 10

[156] I. K. Rakkolainen and K. Palovuori. Interactive digital FogScreen. In *Proceedings of the 3rd Nordic Conference on Human-Computer Interaction*, NordiCHI '04, pages 459–460, New York, NY, USA, 2004. ACM. DOI: 10.1145/1028014.1028096. 10

[157] I. K. Rakkolainen and K. Palovuori. Laser scanning for the interactive walk-through FogScreen. In *Proceedings of the ACM Symposium on Virtual Reality Software and Technology*, VRST '05, pages 224–226, New York, NY, USA, 2005. ACM. DOI: 10.1145/1101616.1101661. 10

[158] F. T. R&D. Wearable communications: Optical fibres. http://www.studio-creatif.com/Gb/Vet/Vet02Prototypes05Fr.htm [Last accessed: November 2012], 2003. 12

[159] F. Redhead and M. Brereton. Getting to the nub of neighbourhood interaction. In *Proceedings of the Tenth Anniversary Conference on Participatory Design 2008*, PDC '08, pages 270–273, Indianapolis, IN, USA, 2008. Indiana University. DOI: 10.1007/978-3-642-03658-3_49. 18

[160] F. Redhead and M. Brereton. Designing Interaction for Local Communications: An Urban Screen Study. In *Proceedings of the 12th IFIP TC 13 International Conference on Human-Computer Interaction: Part II*, INTERACT'09, pages 457–460, Berlin, Heidelberg, 2009. Springer-Verlag. DOI: 10.1007/978-3-642-03658-3_49. 18, 70

[161] R. Rodenstein. Employing the periphery: the window as interface. In *Extended Abstracts on Human Factors in Computing Systems*, CHI '99, pages 204–205, New York, NY, USA, May 1999. ACM. DOI: 10.1145/632716.632844. 10

[162] Y. Rogers. Moving on from Weiser's Vision of Calm Computing: Engaging Ubicomp Experiences. In *Proceedings of the 8th international Conference on Ubiquitous Computing*, UbiComp'06, pages 404–421, Berlin, Heidelberg, 2006. Springer-Verlag. DOI: 10.1007/11853565_24. 37

[163] Y. Rogers, W. R. Hazlewood, P. Marshall, N. Dalton, and S. Hertrich. Ambient Influence: Can Twinkly Lights Lure and Abstract Representations Trigger Behavioral Change? In *Proceedings of the 12th ACM international Conference on Ubiquitous Computing*, Ubicomp '10, pages 261–270, New York, NY, USA, 2010. ACM. DOI: 10.1145/1864349.1864372. 10, 19, 70

[164] E. Rubegni, N. Memarovic, and M. Langheinrich. Talking to strangers: Using large public displays to facilitate social interaction. In *Proceedings of the 14th International Conference on Human-Computer Interaction*, HCI '11, pages 195–204. Springer-Verlag, July 2011. DOI: 10.1007/978-3-642-21708-1_23. 17

[165] E. Rukzio, M. Mueller, and R. Hardy. Design, implementation and evaluation of a novel public display for pedestrian navigation: the rotating compass. In *Proceedings of the 27th International Conference on Human factors in computing systems*, CHI '09, pages 113–122, New York, NY, USA, 2009. ACM. DOI: 10.1145/1518701.1518722. 54

[166] M. Satyanarayanan, P. Bahl, R. Caceres, and N. Davies. The case for VM-based cloudlets in mobile computing. *Internet Computing, IEEE*, 8(4), 2009. DOI: 10.1109/MPRV.2009.82. 53

[167] N. Sawhney, S. Wheeler, and C. Schmandt. Aware community portals: Shared information appliances for transitional spaces. *Personal Ubiquitous Computing*, 5(1):66–70, January 2001. DOI: 10.1007/s007790170034. 14

[168] D. Schmidt, F. Chehimi, E. Rukzio, and H. Gellersen. PhoneTouch: A Technique for Direct Phone Interaction on Surfaces. In *Proceedings of the 23nd Annual ACM Symposium on User Interface Software and Technology*, UIST'10, pages 13–16, New York, NY, USA, 2010. ACM. DOI: 10.1145/1866029.1866034. 71

[169] J. Schrammel, E. Mattheiss, S. Döbelt, L. Paletta, A. Almer, and M. Tscheligi. Attentional Behavior of Users on the Move Towards Pervasive Advertising Media. In J. Müller, F. Alt, and D. Michelis,

editors, *Pervasive Advertising*. Springer Limited London, 2011. DOI: 10.1007/978-0-85729-352-7_14. 74

[170]   SeeSaw Networks, Inc. Place-based digital video advertising: SeeSaw networks. http://www.seesawnetworks.com/. 63

[171]   J. Segen and S. Kumar. Shadow gestures: 3d hand pose estimation using a single camera. In *Computer Vision and Pattern Recognition, 1999. IEEE Computer Society Conference on.*, volume 1. IEEE, 1999. DOI: 10.1109/CVPR.1999.786981. 50

[172]   J. She, J. Crowcroft, H. Fu, and P.-H. Ho. Smart signage: A draggable cyber-physical broadcast/multicast media system. In *Proceedings of the IEEE International Conference on Cyber, Physical and Social Computing*, CPSCom 2012, 2012. DOI: 10.1109/GreenCom.2012.55. 54

[173]   A. S. Shirazi, C. Winkler, and A. Schmidt. Flashlight interaction: a study on mobile phone interaction techniques with large displays. In *Proceedings of the 11th International Conference on Human-Computer Interaction with Mobile Devices and Services*, MobileHCI '09, pages 93:1–93:2, New York, NY, USA, 2009. ACM. DOI: 10.1145/1613858.1613965. 53

[174]   B. Shneiderman. Designing for Fun: How Can We Design User Interfaces to be More Fun? *Interactions*, 11(5):48–50, Sept. 2004. DOI: 10.1145/1015530.1015552. 40

[175]   G. Shoemaker, A. Tang, and K. S. Booth. Shadow reaching: a new perspective on interaction for large displays. In *Proceedings of the 20th annual ACM symposium on User interface software and technology*, UIST '07, pages 53–56, New York, NY, USA, 2007. ACM. DOI: 10.1145/1294211.1294221. 39

[176]   G. B. D. Shoemaker and K. M. Inkpen. Single display privacyware: Augmenting public displays with private information. In *Proceedings of the SIGCHI Conference on Human Factors in Computing Systems*, CHI '01, pages 522–529, New York, NY, USA, March 2001. ACM. DOI: 10.1145/365024.365349. 54, 71

[177]   SMART Technologies. Smart board interactive whiteboards. http://www.smarttech.com/Smart Board [Last accessed: January 2013], 2013. 14, 15

[178]   D. Snowdon and A. Grasso. Diffusing information in organizational settings: Learning from experience. In *Proceedings of the SIGCHI Conference on Human Factors in Computing Systems*, CHI '02, pages 331–338, New York, NY, USA, 2002. ACM. DOI: 10.1145/503376.503435. 14

[179]   R. Sodhi, H. Benko, and A. Wilson. Lightguide: projected visualizations for hand movement guidance. In *Proceedings of the SIGCHI Conference on Human Factors in Computing Systems*, CHI '12, pages 179–188, New York, NY, USA, 2012. ACM. DOI: 10.1145/2207676.2207702. 50

[180]   Sony. Ziris application suite. http://www.sony.co.uk/pro/product/arenasentrydigisign/ziris-application-suite/overview [Last accessed: September 2012], 2004–2012. 59

[181]   L. Story. Anywhere the eye can see, it's likely to see an ad. *The New York Times*, January 15th 2007. 22

[182] O. Storz, A. Friday, and N. Davies. Supporting content scheduling on situated public displays. *Computers & Graphics*, 30(5):681–691, 2006. DOI: 10.1016/j.cag.2006.07.002. 79

[183] O. Storz, A. Friday, N. Davies, J. Finney, C. Sas, and J. Sheridan. Public ubiquitous computing systems: Lessons from the e-Campus display deployments. *Pervasive Computing, IEEE*, 3(5):40–47, 2006. DOI: 10.1109/MPRV.2006.56. 74, 75, 76

[184] N. Streitz, C. Röcker, T. Prante, R. Stenzel, and D. van Alphen. Situated Interaction With Ambient Information: Facilitating Awareness and Communication in Ubiquitous Work Environments. In *Proceedings of the 10th International Conference on Human–Computer Interaction*, HCI International'03, 2003. 34

[185] Ströer Digital Media GmbH. Infoscreen. http://www.stroeerdigital.de/en/infoscreen/ [Last accessed: January 2013], 2013. 17

[186] C. Taivan, R. José, and I. Elhart. Selection and control of applications in pervasive displays. In *6th International Conference on Ubiquitous Computing and Ambient Intelligence (UCAmI 2012)*, December 2012. DOI: 10.1007/978-3-642-35377-2_23. 61, 62

[187] N. Taylor and K. Cheverst. Social interaction around a rural community photo display. *International Journal of Human-Computer Studies*, 67(12), December 2009. DOI: 10.1016/j.ijhcs.2009.07.006. 18

[188] N. Taylor and K. Cheverst. Creating a rural community display with local engagement. In *Proceedings of the 8th ACM Conference on Designing Interactive Systems*, DIS '10, pages 218–227, New York, NY, USA, 2010. ACM. DOI: 10.1145/1858171.1858209. 18

[189] N. Taylor and K. Cheverst. Supporting community awareness with interactive displays. *Computer, IEEE*, 45(5):26–32, 2012. DOI: 10.1109/MC.2012.113. 18

[190] N. Taylor, K. Cheverst, D. Fitton, N. J. P. Race, M. Rouncefield, and C. Graham. Probing communities: Study of a village photo display. In *Proceedings of the 2007 Australasian Computer-Human Interaction Conference*, OzCHI '07. ACM, November 2007. DOI: 10.1145/1324892.1324896. 18

[191] M. Ten Koppel, G. Bailly, J. Müller, and R. Walter. Chained displays: configurations of public displays can be used to influence actor-, audience-, and passer-by behavior. In *Proceedings of the SIGCHI Conference on Human Factors in Computing Systems*, CHI '12, pages 317–326, New York, NY, USA, 2012. ACM. DOI: 10.1145/2207676.2207720. 48

[192] D. Vogel and R. Balakrishnan. Interactive public ambient displays: Transitioning from implicit to explicit, public to personal, interaction with multiple users. In *Proceedings of the 17th Annual ACM symposium on User Interface Software and Technology*, UIST '04, pages 137–146, New York, NY, USA, 2004. ACM. DOI: 10.1145/1029632.1029656. 34, 50

[193] R. Walter, G. Bailly, and J. Müller. Strikeapose: Revealing mid-air gestures on public displays. In *Proceedings of the 2013 ACM Conference on Human Factors in Computing Systems*, CHI'13, New York, NY, USA, 2013. ACM. DOI: 10.1145/2470654.2470774. 49, 50, 51

[194] M. Wang, S. Boring, and S. Greenberg. Proxemic peddler: A public advertising display that captures and preserves the attention of a passerby. In *Proceedings of the 2012 International Symposium on Pervasive Displays*, PerDis '12, New York, NY, USA, 2012. ACM. DOI: http://dx.doi.org/10.1145/2307798.2307801. 35

[195] R. Want, A. Hopper, V. F. ao, and J. Gibbons. The active badge location system. *ACM Transactions on Information Systems*, 10(1):91–102, January 1992. DOI: 10.1145/128756.128759. 8

[196] M. Weiser. The computer for the 21st century. *Scientific American*, September 1991. DOI: 10.1038/scientificamerican0991-94. 8

[197] M. Weiser and J. S. Brown. The Coming Age of Calm Technolgy. In P. J. Denning and R. M. Metcalfe, editors, *Beyond Calculation*, pages 75–85, New York, NY, USA, 1997. Copernicus. DOI: 10.1007/978-1-4612-0685-9_6. 9, 37

[198] Wi-Fi Alliance. Wi-Fi CERTIFIED Miracast™. http://www.wi-fi.org/miracast [Last accessed: October 2012], 2012. 53

[199] D. Wigdor and D. Wixon. *Brave NUI world: designing natural user interfaces for touch and gesture*. Morgan Kaufmann, 2011. 49

[200] A. Williams, S. D. Farnham, and S. Counts. Exploring wearable ambient displays for social awareness. In *Extended Abstracts on Human Factors in Computing Systems*, CHI '06, pages 1529–1534, New York, NY, USA, April 2006. ACM. DOI: 10.1145/1125451.1125731. 11

[201] S. Wilson, J. Galliers, and J. Fone. Not all sharing is equal: The impact of a large display on small group collaborative work. In *Proceedings of the 2006 20th Anniversary Conference on Computer Supported Cooperative Work*, CSCW '06, pages 25–28, New York, NY, USA, 2006. ACM. DOI: 10.1145/1180875.1180880. 15

[202] A. Wolbach, J. Harkes, S. Chellappa, and M. Satyanarayanan. Transient customization of mobile computing infrastructure. In *Proceedings of the First Workshop on Virtualization in Mobile Computing*, MobiVirt '08, New York, NY, USA, June 2008. ACM. DOI: 10.1145/1622103.1622108. 31, 53

[203] Xerox Corporation. Flowport, workflow software: Xerox. http://www.xerox.com/digital-printing/workflow/printing-software/flowport/enus.html [Last accessed: January 2013], 2013. 14

Authors' Biographies

NIGEL DAVIES

Nigel Davies is a Professor in the School of Computing and Communications at Lancaster University and is currently a visiting faculty member at CMU. His research focuses on experimental mobile and ubiquitous systems and his projects include the MOST, GUIDE and e-Campus projects that have been widely reported on in the academic literature and the popular press. Professor Davies has held visiting positions at SICS, Sony's Distributed Systems Lab in San Jose, the Bonn Institute of Technology, ETH Zurich and Google Research in Mountain View, CA. Nigel is active in the research community and has co-chaired both Ubicomp and MobiSys conferences. He is currently editor-in-chief of *IEEE Pervasive Magazine,* chair of the steering committee for Hot-Mobile and one of the founders of the ACM PerDis Symposium on Pervasive Displays. From 2010 to 2013 he was the coordinator of PD-NET, a large multinational project that aimed to lay the scientific foundations for the next generation of open pervasive display networks.

SARAH CLINCH

Sarah Clinch is a PhD researcher at Lancaster University, UK, where she works on supporting applications and personalization in pervasive display systems. In 2009 she was a visiting scholar at Carnegie Mellon University where she conducted research on cloudlet computing and the use of virtualization techniques for public displays. Sarah also worked on the PD-NET project for future display networks in collaboration with the University of Stuttgart, the University of Minho and the University of Lugano. She has published extensively in the field and is a co-creator of the Yarely digital signage software system that is in daily use at Lancaster. Sarah has served as the publicity chair for PerDis 2012 and is the Web Chair for Ubicomp 2014.

FLORIAN ALT

Florian Alt is an Assistant Professor in the Group for Media Informatics at the University of Munich, Germany. His research interest is on pervasive advertising and on implicit and explicit interaction with public displays. Particularly he looks at how to entice people to interact and investigates the cognitive effects of interaction. Florian's research projects include Digifieds, a digital public notice area that allowed users' expectations and suitable interaction techniques to be evaluated, and Looking Glass, an interactive gesture-based display application investigating how interactivity can be communicated to passersby. His research has been published in leading HCI venues and journals (SIGCHI, ToCHI) and he is a co-editor of the book *Pervasive Advertising* that appeared in the Springer HCI Series in 2011. Prior to his appointment in Munich, Florian worked as a research associate at the University of Stuttgart and as a visiting researcher at Telekom Innovation Labs in Berlin. He holds a diploma in media informatics from the University of Munich and a Ph.D. in computer science from the University of Stuttgart.

Printed in the United States
by Baker & Taylor Publisher Services